AF576029

Future Challenges in Rabbit Nutrition

Future Challenges in Rabbit Nutrition

Editors

Francesco Gai
Laura Gasco
Angela Trocino

MDPI • Basel • Beijing • Wuhan • Barcelona • Belgrade • Manchester • Tokyo • Cluj • Tianjin

Editors
Francesco Gai
Institute of Sciences of Food Productions,
National Research Council
Italy

Laura Gasco
Department of Agricultural, Forest and Food
Sciences, University of Turin
Italy

Angela Trocino
Department of Comparative Biomedicine and Food Sciences,
University of Padova
Italy

Editorial Office
MDPI
St. Alban-Anlage 66
4052 Basel, Switzerland

This is a reprint of articles from the Special Issue published online in the open access journal *Animals* (ISSN 2076-2615) (available at: https://www.mdpi.com/journal/animals/special_issues/Future_challenges_in_Rabbit_Nutrition#).

For citation purposes, cite each article independently as indicated on the article page online and as indicated below:

LastName, A.A.; LastName, B.B.; LastName, C.C. Article Title. *Journal Name* **Year**, *Volume Number*, Page Range.

ISBN 978-3-0365-0954-9 (Hbk)
ISBN 978-3-0365-0955-6 (PDF)

Contents

About the Editors

Francesco Gai MSc in Biology (2000) and PhD in Animal Productions (2006) at University of Turin (NW Italy). Senior researcher at the Institute of Sciences of Food Production (ISPA), National Research Council (CNR), Grugliasco (Italy). Expertise in animal feeding and nutritional strategies with the aim to improve quality, safety, and sustainability of food of animal origin. He's co-author of 116 manuscripts in peer reviewed journals (h-index 28), with 2486 citations and 12 book chapters. Membership of editorial boards of international scientific journals: Editorial Board Animals https://www.mdpi.com/journal/animals/sectioneditors/Animal_Nutrition External evaluator of H2020 research and innovation proposals for European Commission Research Executive Agency (2019); External evaluator of grant applications for National Science Centre of Poland (2015, 2018); External evaluator of grant applications for Croatian Science Foundation (2019); External evaluator of grant applications for French National Research Agency (2019); External Reviewer for Enterprise Ireland (CAREER-FIT MSCA COFUND Career Development Fellowships in Ireland's Technology Centre Programme) Horizon 2020 Marie Curie Career-FIT Applications Dublin, Ireland; Peer Reviewer of 136 publications for 49 international journals (2005–2020): https://publons.com/researcher/1560271/francesco-gai/peer-review.

Laura Gasco PhD, Full Professor at Department of Agricultural, Forest, and Food Sciences, University of Turin, Italy. Agricultural engineer (1992) and PhD in Animal Sciences (1999). More than 25 years of experience in Animal Sciences (Scopus ID: 12807274200; h-index: 35). Research topics: fish, rabbits and poultry rearing, nutrition and product quality. Since 2012, she has been involved in trials on insects as raw material in fish and poultry nutrition, and in trials on insect rearing for feed purposes. Scientific Coordinator and Partner of the following projects: CARIPLO: Circular economy: live larvae recycling organic waste as sustainable feed for rural poultry (CELLOW-FeeP) (Partner); AQUAEXCEL3.0: AQUAculture infrastructures for EXCELlence in European fish research 3.0. Coordinator of the UNITO (DISAFA) for the insect and aquaculture platform. SUSINCHAIN (2019): SUStainable INsect CHAIN. Horizon 2020 LC-SFS-17-2019). WP4 (insects as feed) leader (Proposal number: 861976); WELFINSECTS (2019): FROM WASTE TO FARM Insect larvae as tool for welfare improvement in poultry—EU project EIT-FOOD (Partner), Aquaexcel Programme SanHer (AE070026-2018) "Insect meal in pikeperch diets" (Coordinator) INNOPOULTRY (2018): The poultry food chain: tackling old problems with innovative approaches - EU project EIT-FOOD (Patner); HI-BIORAFINERY (2018) Pilot plan of black soldier fly reared on large scale (Polo Innovazione Regionale: Energy and Clean Technologies) (Partner) 2017Aquaexcel Programme-AIRE PROGRAMM—AE040069. Effects of honey bee pollen on growth performance and immune hematological profile of meagre (Argyrosomus regius) Aquaexcel Programme—IMPRovES—AE040040. Use on pre-treated insect meals in European sea bass (Dicentrarchus labrax) to improve nutrient availability and performance; 2016 AGER: Fine Feed For Fish (4F) (Coordinator Research Unit); Blue Economy Challenge (http://theblueeconomychallenge.com/announcingthe-winners-of-the-blue-economy-challenge/): Larvae from biowaste for aquaculture feed (LAWAF) (Partner).

Angela Trocino Associate professor at the Department of Comparative Biomedicine and Food Science, University of Padova, Italy. PhD in Animal Science. Coordinator of the PhD Course in Animal and Food Science, University of Padova. Scientific responsible of Project Unimpresa (2019), PhD project Unismart (2019), Marie Curie PISCOPIA (2014), Project FEAMP (2017), Project of the University of Padova (2014), Project CNR-Agenzia2000 young researchers; responsible for a research unit in project PSR-ISAL (2014). Participant to H2020-RUR-2018-2020 NETPOULSAFE (2020–2023), PRIMA FISH-PhotoCAT (2020–2023), Concerted action ERAFE (FAIR3-CT96-1651) (1996–2002); Expert COST Action 848 (2000–2005); national research projects granted by the Italian Ministry of University and Research (MIUR) and Ministry of Agriculture, Food & Forestry. In collaboration with Italian and foreign researchers, her research activity deals with: rearing systems and welfare of rabbits, broiler chickens and layer hens; harmonization of studies in rabbit nutrition and feeding; protein and energy metabolism, nutritional requirements and relationships with health in growing rabbits and reproducing does. Among the other research topics, effects of feeding, rearing, and housing systems on meat quality (veal, rabbit, fish, broiler chickens); effect of rearing systems of welfare, freshness, quality, and safety of fish and molluscs. Author and co-author of more than 200 original documents, including research papers, communications at congresses and book chapters. In Scopus, 86 publications and H-index 22.

Preface to "Future Challenges in Rabbit Nutrition"

Rabbit breeding, although being a small sector of animal husbandry, is widespread in many areas of the world as the rabbit is intended both for food (meat) and not food (fur) purposes. Rabbit production chain has to face various problems, mainly concerning animal health and product quality. Gastrointestinal disorders and epizootic rabbit enteropathy are major health problems in rabbit production, causing high mortality rates around and after weaning, which also affects global farm efficiency. Considering the European commitment to responsible antibiotic use in animals, rabbit nutrition and feeding play a very important role in contrasting these problems. Besides, the increasing cost of conventional feed ingredients and the increasing interest towards circular economy push research to look for alternative raw materials able to provide new sources of proteins, energy or other valuable components for rabbit diets with a sustainable environmental impact.

To overcome these issues, studies using a multidisciplinary approach addressing aspects of the rabbit nutrition and feeding, with a direct impact on the rabbit farming, welfare, health, and meat quality are particularly appreciated and requested by the scientific community.

This Special Issue received four original papers and one review focused on different nutritional approaches. In particular a phyto-additive (thyme essential oil) and a rabbit-derived bacteriocin-producing strain (*Enterococcus faecium* CCM7420) with probiotic properties were investigated as new feed additives while two types of insect fats and their effects as dietary replacement of soybean oil and their in vitro antimicrobial activities were studied as alternative raw materials. Results collected in this Special Issue will be of particular interest for farmers and animal nutritionists working in the rabbit.

Francesco Gai, Laura Gasco, Angela Trocino
Editors

Article

Effect of Dietary Insoluble and Soluble Fibre on Growth Performance, Digestibility, and Nitrogen, Energy, and Mineral Retention Efficiency in Growing Rabbits

Carlos Farías-Kovac [1], Nuria Nicodemus [1], Rebeca Delgado [1], César Ocasio-Vega [1], Tamia Noboa [1], Ramadan Allam-Sayed Abdelrasoul [2], Rosa Carabaño [1] and Javier García [1,*]

1 Departamento de Producción Agraria, Escuela Técnica Superior de Ingeniería Agronómica, Agroalimentaria y de Biosistemas, Universidad Politécnica de Madrid, C/Senda del Rey 18, 28040 Madrid, Spain; scarlos_03@hotmail.com (C.F.-K.); nuria.nicodemus@upm.es (N.N.); rebeca_dm85@yahoo.es (R.D.); cesar.ocasio1@gmail.com (C.O.-V.); taminobo31@hotmail.com (T.N.); rosa.carabano@upm.es (R.C.)

2 Poultry Department, Faculty of Agriculture, Fayoum University, Fayoum 63514, Egypt; ras05@fayoum.edu.eg

* Correspondence: javier.garcia@upm.es; Tel.: +34-9106-71037; Fax: +34-91-549-97-63

Received: 15 July 2020; Accepted: 31 July 2020; Published: 4 August 2020

Simple Summary: Rabbits, like other herbivores, require a minimal level of insoluble fibre in the diet to warrant an adequate digestive function. Epizootic rabbit enteropathy (ERE) is the main digestive trouble in growing rabbits that increases the use of antibiotics. The increase of soluble fibre (once met insoluble fibre requirements) limits the incidence of ERE and improves nitrogen and energy balance, while high levels of insoluble fibre seem to favour ERE. This study evaluated whether the increase of soluble and insoluble fibre above the current requirements of insoluble fibre had a positive impact on mortality, growth performance, diet digestibility, and energy, nitrogen, and mineral balance. Treatments had no effect on mortality, which was low (1%). The increase of insoluble fibre reduced the dietary digestible energy, while soluble fibre only increased it when combined with a low insoluble fibre level. The group fed with the lowest insoluble and soluble fibre levels showed the best energy and mineral balance, while the increase of insoluble or soluble fibre did not improve any growth trait. We conclude that in healthy rabbits, and once the minimal insoluble fibre requirements are met, no increase of insoluble or soluble fibre is recommended.

Abstract: Dietary soluble fibre limits the incidence of epizootic rabbit enteropathy (ERE) and improves the energy and nitrogen balance in low-insoluble fibre diets, while high-insoluble fibre diets seem to favour ERE. This study assessed whether the positive effects of soluble fibre are influenced by the level of insoluble fibre. Four diets (2 × 2 factorial arrangement) were used with two levels of insoluble fibre (314 vs. 393 g/kg DM) and soluble fibre (87 vs. 128 g/kg DM), resulting in four diets with increasing total dietary fibre levels. Growth performance and chemical composition (body and carcass) (28–62 days of age), faecal digestibility (54–57 days of age), and jejunal morphometry functionality (39 days of age) were determined. Mortality was low (<1%) and treatments did not influence it. Insoluble and soluble fibre tended to reduce the growth rate ($p \leq 0.109$), body protein, and fat accretion ($p = 0.049$ to 0.120), but only insoluble fibre impaired feed efficiency ($p < 0.001$). The efficiency of digestible energy used for growth was impaired with the increase of total dietary fibre ($p = 0.027$), while that of nitrogen remained majorly unaffected. In conclusion, in healthy rabbits, the increase of either insoluble or soluble fibre had no benefit.

Keywords: insoluble fibre; soluble fibre; feed efficiency; whole body and carcass chemical composition; energy nitrogen and mineral balance; fibre digestibility; mucosa morphology; energy nitrogen and mineral retention efficiency; rabbit

---

1. Introduction

The inclusion of a moderated level of soluble fibre (~12%; SF), usually derived from sugar beet pulp (SBP) inclusion, enhanced gut functionality [1,2] and modified intestinal microbiota [3–5], which was related to a decrease of the mortality rate in rabbits affected by epizootic rabbit enteropathy (ERE) [6–8]. These positive effects associated with SBP were accounted not only for its SF but also for its high content of easily fermentable insoluble fibre [9].

The inclusion of around 15–25% of SBP is required to meet the recommended SF level. However, the energy and nutrient retention efficiency in growing rabbits seem to be directly affected by the level of fermentable fibre. When SBP was included above 15% in the diet replacing barley, it reduced the energy and nitrogen retention efficiency [10]. Nevertheless, a 30% SBP level of inclusion, in substitution of barley and alfalfa, did not affect these traits in low-fibrous diets [11]. On the opposite, energy and nitrogen retention efficiency increased when a lower level of SBP was used (18%) in low-fibre diets (31% neutral detergent fibre, NDF, on DM basis and free of ash and protein) [8], probably due to the improvement of the rabbit health status and the reduction of urine nitrogen losses. In this context, the increase of the insoluble fibre level (NDF) impaired the nutritive value of SBP [12], which might be related with the role of NDF in modulating the microbial activity and rate of passage [13,14], and finally might affect the dietary energy and nitrogen efficiency. The use of high-insoluble fibrous diets for fattening rabbits is not strange at the field level, even when they might increase the incidence of ERE [15,16].

The aim of this work was to clarify whether the positive effect of the moderated inclusion of SF is influenced or not by the NDF level on growth performance, diet digestibility, gut mucosa morphometry and functionality, and energy, protein and mineral retention efficiency of growing rabbits.

2. Materials and Methods

2.1. Animals and Housing

A total of 264 crossbred hybrid rabbits (New Zealand White × Californian, V × R line from UPV, Valencia, Spain) weaned at 28 days of age were used. They came from 71 multiparous rabbit does from a farm periodically affected by ERE. Rabbits after weaning received the same diets as those offered to their mothers (described below). Rabbits were individually caged in flat deck cages (610 × 250 × 330 mm), except during the faecal digestibility trial in which they were kept in wire metabolism cages (405 × 510 × 320 mm) that allowed the separation of faeces and urine. Rabbits had *ad libitum* access to feed and water with no antibiotic supplemented throughout the experimental period and their health was checked daily. The temperature ranged between 18 and 23 °C. All the experimental procedures used were in compliance with the Spanish guidelines for care and use of animals in research [17], and authorized by the Dirección General de Agricultura y Ganadería from the Community of Madrid (PROEX 328/15).

2.2. Diets

Four diets in a 2 × 2 factorial arrangement were used with two levels of insoluble fibre quantified as NDF (314 vs. 393 g/kg DM, free of ash and protein) and two levels of SF quantified by the difference of total dietary fibre and NDF (87 vs. 128 g/kg DM) (Table 1). A control diet was formulated to meet most nutrient requirements for growing rabbits [18], with an NDF level (344 g NDF free of ash/kg DM) similar to that proposed for post-weaned rabbits [15] (342 g NDF free of ash/kg DM), and with a level of SF lower than the one recommended for growing rabbits under ERE risk [7] (LIF-LSF diet). The increase of NDF was obtained by replacing barley and wheat from the LIF-LSF diet for wheat straw, defatted grape seed meal, dehydrated alfalfa, and lard, the latter to minimize the reduction of digestible energy (DE) content (HIF-LSF diet). The increase of SF was obtained by replacing wheat bran, sunflower meal, gluten feed, wheat straw, and grape seed meal from the LIF-LSF diet for SBP (LIF-HSF diet). A fourth diet high in both NDF and SF was formulated (HIF-HSF diet).

Table 1. Ingredients and chemical composition of the experimental diets.

Item [1]	Low-Insoluble Fibre		High-Insoluble Fibre	
	Low SF	High SF	Low SF	High SF
	LIF-LSF	LIF-HSF	HIF-LSF	HIF-HSF
Ingredient, g/kg as fed				
Barley	150.0	141.0	70.0	70.0
Wheat	150.0	141.0	70.0	70.0
Dehydrated alfalfa	101.8	101.8	154.0	154.0
Cereal straw	84.0	63.0	169.3	146.6
Defatted grape seed meal	36.0	27.0	72.6	62.9
Wheat bran	176.5	60.0	133.8	10.0
Sugar beet pulp	0.0	170.0	0.0	170.0
Gluten feed	93.5	30.5	71.3	5.0
Sunflower meal	110.0	171.0	130.0	185.0
Soybean meal	60.0	60.0	60.0	60.0
Lard	0.0	0.0	35.0	35.0
Cane molasses	10.0	10.0	10.0	10.0
L-Lysine 50	3.0	2.4	3.2	2.6
Alimet	0.5	0.5	0.9	0.9
L-Threonine	1.0	0.6	1.2	0.8
L-Tryptophan	0.2	0.2	0.2	0.2
Sodium chloride	2.5	2.5	2.5	2.5
Calcium carbonate	16.0	8.0	10.0	4.0
Calcium phosphate	3.5	9.0	4.5	9.0
Mineral/vitamin premix [1]	1.5	1.5	1.5	1.5
Chemical composition, g/kg DM				
DM	903	905	908	906
Ash	83.9	89.1	90.9	96.3
Total dietary fibre [2]	398	444	484	519
Neutral detergent fibre (NDF [2])	311	317	396	390
Acid detergent fibre [3]	162	179	230	247
Acid detergent lignin [3]	36.6	38.3	66.4	65.2
Soluble fibre [4]	86.7	127.5	88.1	129.0
Crude protein (CP)	197	198	187	183
CP-Total dietary fibre	61.1	67.9	69.2	72.3
CP-NDF	33.8	37.4	33.8	38.0
Starch	212	181	117	86.3
Ether extract	31.6	35.9	79.3	65.7
Gross energy, MJ/kg DM	18.0	17.7	18.9	18.3

[1] Mineral and vitamin composition (per kg of complete diet): 20.0 mg of Mn; 60 mg of Zn; 12 mg of Cu; 1.01 mg of I; 0.30 mg of Co; 42 mg of Fe; 0.12 mg of Se; 9750 UI of vitamin A; 1200 UI of vitamin D_3, 42 UI of vitamin E, 1.5 mg of vitamin K_3; 1.5 mg of vitamin B_1; 3.75 mg of vitamin B_2; 1.5 mg of vitamin B_6; 0.018 mg of vitamin B_{12}; 12 g of Pantothenic acid; 40 g of Nicotinic acid; 0.75 mg of Folic acid; 0.075 mg of Biotin. [2] Corrected for ash and protein. [3] Corrected for ash. [4] Quantified as total dietary fibre NDF (both corrected for ash and protein).

2.3. Growth Performance and Body Chemical Composition Trial

In total, 224 rabbits (56/diet) weighing 510 ± 74 g were weaned at 28 days of age, and growth rate, feed intake, feed efficiency, and mortality was recorded until 62 days of age. The whole body and carcass chemical composition and energy content was estimated in vivo using the bioelectrical impedance analysis (BIA) technique in 39 rabbits/diet (randomly selected from this group), weighing 511 ± 77 g.

Measurements of resistance and reactance were measured in rabbits with a body composition analyser (Model Quantum II, RJL Systems, Detroit, MI, USA) at 28 and 62 days of age [19,20]. Multiple regression equations were used to estimate water, protein, ash, fat, and energy proportions both in the whole body and in the carcass according to these authors. Twenty-four rabbits were discarded due to an excess of feed waste (5, 2, 9, and 8 rabbits from the LIF-LSF, LIF-HSF, HIF-LSF, and HIF-HSF groups, respectively) and two rabbits died from the LIF-HSF group.

2.4. Faecal Digestibility Trial

A group of 40 rabbits (10/diet) weighing 1905 ± 108 g (belonging to the growth performance group) were caged individually in metabolism cages to determine the apparent faecal digestibility of gross energy, protein, and fibre fractions according to Perez et al. [21]. Feed intake and total faecal output were recorded for each rabbit during four consecutive days (from 54 to 57 days of age). Faeces were collected daily and were stored at −20 °C, dried at 80 °C for 48 h, and ground with a 1-mm screen for analysis.

2.5. Calculations of Energy and Nitrogen Efficiency

Estimated values for the total whole-body nitrogen and energy content were used to obtain the nitrogen and energy retention in the whole body (NR whole body and ER whole body, respectively) at 28 and 62 days of age. Values were expressed per kg $BW^{0.75}$ and day (where the BW was calculated as the average of the final and the initial body weight). Estimated values for the carcass nitrogen, and energy content were used to calculate the nitrogen and energy retention in the carcass (NR carcass and ER carcass, respectively) at 28 and 62 days of age. Values were also expressed per kg $BW^{0.75}$ and day. Moreover, nitrogen and gross energy intake (Ni and GEi, respectively) and digestible N and DE intake (DNi and DEi, respectively) were recorded to calculate the overall N and GE whole-body retention efficiency as: NR whole body/Ni, NR whole body/DNi, ER whole body/GEi and ER whole body/DEi. Besides, the overall N and GE carcass retention efficiency was calculated as: NR carcass/DNi and ER carcass/DEi.

Total N and GE excretion as skin and viscera, faeces, or heat production and urine were calculated as follows:

$$\text{N lost as skin and viscera (g/kg BW}^{0.75}\text{ and day)} = \text{(g NR in the whole body} - \text{g NR in the carcass)/kg BW}^{0.75}\text{ and day.} \quad (1)$$

$$\text{N excreted as faeces (g/kg BW}^{0.75}\text{ and day)} = \text{(Ni} - \text{DNi)/kg BW}^{0.75}\text{ and day.} \quad (2)$$

$$\text{N excreted as urine (g/kg BW}^{0.75}\text{ and day)} = \text{(DNi} - \text{NR in the whole body)/kg BW}^{0.75}\text{ and day.} \quad (3)$$

$$\text{Energy lost as skin and viscera (MJ/kg BW}^{0.75}\text{ and day)} = \text{(MJ ER in the whole body} - \text{MJ ER in the carcass)/kg BW}^{0.75}\text{ and day.} \quad (4)$$

$$\text{Energy excreted as faeces (MJ/kg BW}^{0.75}\text{ and day)} = \text{(GEi} - \text{DEi)/kg BW}^{0.75}\text{ and day.} \quad (5)$$

$$\text{Energy excreted as urine and heat production (MJ/kg BW}^{0.75}\text{ and day)} = \text{(DEi} - \text{RE in the whole body)/kg BW}^{0.75}\text{ and day.} \quad (6)$$

Energy retained as protein and fat was calculated taking into account that energy deposited as protein and fat equals 23.15 and 35.65 kJ/g [22]. Energy retained as fat was also calculated as the difference between the total retained energy and the retained energy as protein. Digestible nitrogen and DE intake used for production was obtained by the difference between the total DNi and DEi and the nitrogen and energy requirements for maintenance, respectively (0.464 g DN/kg $BW^{0.75}$ day, and 430 kJ DE/kg $BW^{0.75}$ day) [22]. Mineral balance was estimated in the same way as the nitrogenous and energetic ones.

2.6. Gut Histology and Enzymatic Activity and Immune Function

Another different group of 40 rabbits (10/treatment) of 28 days of age and 472 ± 84 g BW were caged in groups of 2 rabbits (5 cages/treatment) and slaughtered by head concussion at 39 days of age. A sample from the middle part of the jejunum (3 cm) were collected in 10% buffered neutral formaldehyde solution (pH 7.2 to 7.4) for histological analysis and processed as described [23]. Another 6 cm were excised from the middle part of the jejunum, flushed with saline solution, frozen in dry ice, and immediately stored at −80 °C to determine sucrose-specific activity as described [1]. Villous height, crypt depth, and goblet cell from each villous were measured individually. Seven samples were discarded due to poor staining or conservation (3, 1, and 3 from the LIF-HSF, HIF-LSF, and HIF-HSF groups, respectively).

2.7. Chemical Analysis

Procedures of the AOAC [24] were used to determine DM (method 934.01), ash (method 942.05), crude protein (method 968.06), ether extract (920.39), starch (amyloglucosidase-α-amylase method; method 996.11), and total dietary fibre (985.29; TDF). Dietary NDF, acid detergent fibre, and acid detergent lignin were determined sequentially using the filter bag system (Ankom Technology, New York, NY, USA) [24,25]. Dietary NDF was determined using a thermo-stable amylase without any sodium sulphite added and corrected for ash and protein. The dietary SF was calculated, as TDF–NDF (both corrected for ash and protein), and no correction for mucin was done in the faecal samples. Gross energy (GE) was measured by an adiabatic bomb calorimeter (model 356, Parr Instrument Company, Moline, IL, USA).

2.8. Statistical Analysis

Data of faecal digestibility, growth performance, sucrose activity, and nitrogen and energy balances were analysed as a completely randomized design with the level of NDF, SF, and their interaction as the main sources of variation by using the mixed procedure of SAS (SAS Inst., Cary, NC, USA). Weaning weight was used as a linear covariate for growth traits, whole body and carcass chemical composition, and nitrogen, energy and mineral balances. Mortality was analysed using a logistic model (GENMOD procedure of SAS using a binomial distribution) considering the same variables, and the results were transformed from the logit scale. All data were presented as least-squares means. When interactions were significant, comparisons among all the treatment means were made using a *t*-test.

3. Results

In the whole experimental period, rabbits fed with HIF diets showed a higher feed intake (by 11%; $p < 0.001$, Table 2) and tended to decrease the growth rate ($p = 0.109$), leading to a lower feed efficiency (by 12%; $p < 0.001$) compared with the LIF group. In contrast, rabbits fed HSF diets reduced their feed intake (by 2%; $p = 0.048$), with no effect on the feed efficiency. No interaction was observed between NDF and SF in growth traits. Treatments did not affect the mortality rate, which was lower than 1%.

Table 2. Effect of dietary level of insoluble and soluble fibre on rabbits' growth performance from 28 to 62 days of age.

Item	Diets [1]				SEM		p-Value		
	Low-Insoluble Fibre		High-Insoluble Fibre						
	Low-Soluble Fibre LIF-LSF	High-Soluble Fibre LIF-HSF	Low-Soluble Fibre HIF-LSF	High-Soluble Fibre HIF-HSF	IF and SF	IF × SF	IF	SF	IF × SF
N	51	52	47	48					
28–42 days of age									
Body weight 28 day, g	496	459	533	540	6.58	9.30	<0.001	0.103	0.018
Growth rate, g/day	52.5	51.1	52.6	52.0	0.66	0.93	0.660	0.266	0.653
Feed intake, g/day	107	107	118	118	1.89	2.68	<0.001	0.907	0.974
Feed efficiency, g/g	0.519	0.489	0.447	0.445	0.010	0.015	<0.001	0.277	0.340
42–62 days of age									
Body weight 42 day,g	1241	1221	1242	1233	9.172	13.01	0.656	0.263	0.649
Growth rate, g/day	50.7	50.3	49.2	47.9	0.50	0.71	0.012	0.234	0.517
Feed intake, g/day	152	148	170	164	1.51	2.15	<0.001	0.009	0.515
Feed efficiency, g/g	0.336	0.345	0.291	0.294	0.004	0.005	<0.001	0.276	0.581
28–62 days of age									
Body weight 62 g/day	2254	2227	2225	2191	13.39	19.01	0.109	0.098	0.867
Growth rate, g/day	51.4	50.6	50.6	49.6	0.39	0.56	0.109	0.099	0.870
Feed intake, g/day	133	131	149	145	1.22	1.73	<0.001	0.048	0.650
Feed efficiency, g/g	0.388	0.391	0.341	0.344	0.003	0.005	<0.001	0.541	0.968

[1] LIF-LSF = Low-insoluble fibre, Low-soluble fibre; LIF-HSF = Low-insoluble fibre, High-soluble fibre HIF-LSF = High-insoluble fibre, Low-soluble fibre; HIF-HSF = High-insoluble fibre, High-soluble fibre. Weight at weaning (28 days) as a covariate.

The digestibility of dry and organic matter decreased with the NDF level (by 25% and 14%, respectively; $p < 0.001$. Table 3) but increased with the level of SF (by 3.5% and 4%, respectively; $p < 0.001$). In contrast, the ash digestibility decreased with both the level of NDF and SF (by 10% and 8%, respectively; $p \leq 0.014$). An interaction between the level of NDF and SF was observed for the digestibility of energy, protein, total dietary fibre, and NDF ($p \leq 0.037$). Rabbits fed HIF diets had a lower GE and protein digestibility than those fed LIF diets (by 11% and 4%; $p < 0.001$), while the increase of SF improved GE and protein digestibility with LIF but not with HIF diets. The increase of NDF at the low SF level (LIF-LSF vs. HIF-LSF) did not modify the digestibility of any fibre fraction, but the increase of SF improved the digestibility of NDF and SF ($p < 0.001$), being higher in LIF than in HIF groups. These values resulted in very similar ratios of dietary digestible protein/DE among groups, although those of the HSF groups were lower ($p < 0.001$), and that of the HIF-HSF group tended to be the lowest ($p < 0.070$).

At 28 days of age, rabbits from HIF groups showed a higher whole-body fat, and a lower whole-body protein and mineral proportions than those from LIF groups (22.4% vs. 23.4% for fat, 59.1% vs. 58.5% for protein, and 11.7% vs. 11.5% for minerals, all of them on a DM basis; $p < 0.001$. Table S1), with no differences in the carcass protein concentration (Table S2). However, an interaction NDF × SF was found for the fat and mineral proportions, being higher for the LIF-HSF and HIF-LSF groups ($p \leq 0.062$. Table S2). At 62 days of age, there were no differences in the whole-body fat proportion, but the whole-body protein concentration increased in the HIF and HSF groups (in both by 1%; $p \leq 0.032$). Additionally, rabbits fed HSF diets increased in the carcass the protein and decreased the fat proportions (by 0.8 and 3%, respectively; $p \leq 0.028$), and a similar trend was observed for the HIF groups ($p \leq 0.067$), while the HIF diets tended to increase the carcass mineral proportion ($p = 0.055$). The daily whole-body protein and fat accretion tended to reduce with the increase of NDF and SF ($p = 0.049$ to 0.120, Table S3), with no effect on mineral accretion. The composition of the body weight gain had lower fat in the HIF than in the LIF groups ($p = 0.050$), and a lower protein content ($p = 0.034$). The daily carcass fat accretion and the fat proportion in the carcass weight gain decreased in the HSF compared to the LSF groups (by 5% and 2%; $p \leq 0.039$, Table S4), while the daily carcass protein and mineral accretion and the protein and mineral proportions in the carcass decreased in the HIF compared to the LIF groups (by 4% and 2% for protein and by 3 and 1% form minerals; $p \leq 0.060$).

The digestible nitrogen intake was higher for the HIF-LSF group compared with the other three groups (2.41 vs. 2.26 g DNi/kg $BW^{0.75}$ d; $p = 0.003$, Table 4). The nitrogen retained in the whole body and in the carcass decreased by 2–3% when NDF increased ($p \leq 0.040$), with no effect of SF. These results led to the lowest DN retention efficiency in the whole body and in the carcass in the HIF-LSF group compared to the other three groups (0.427 vs. 0.460, and 0.284 vs. 0.310; $p \leq 0.011$), as well as for the retention efficiency of the estimated DN used for growth (0.506 vs. 0.553, and 0.336 vs. 0.372; $p < 0.011$). Faecal nitrogen losses were lower in the LIF groups, and HSF diets reduced them when combined with the LIF diet but increased with the HIF diet ($p < 0.001$). They were inversely proportional to the digestible crude protein content of the diets. Urinary nitrogen losses were the highest for the HIF-LSF group, showing similar values to the other three groups (1.38 vs. 1.23 g urinary/kg $BW^{0.75}$ d; $p = 0.002$).

Table 3. Effect of dietary level of insoluble and soluble fibre on faecal apparent digestibility of dietary components in rabbits from 54 to 57 days of age.

Item	Diets [1]				SEM		*p*-Value		
	Low-Insoluble Fibre		High-Insoluble Fibre						
	Low-Soluble Fibre LIF-LSF	High-Soluble Fibre LIF-HSF	Low-Soluble Fibre HIF-LSF	High-Soluble Fibre HIF-HSF	IF and SF	IF × SF	IF	SF	IF × SF
N	9	8	10	9					
Initial body weight, g	1961	1856	1939	1904	23.26	32.88	0.691	0.042	0.300
Feed intake, g DM/day	135	128	136	142	3.28	4.63	0.113	0.803	0.180
Faecal apparent digestibility, %									
Dry matter	63.9	66.8	55.7	57.0	0.40	0.57	<0.001	<0.001	0.111
Organic matter	65.1	68.5	56.6	58.1	0.44	0.62	<0.001	<0.001	0.147
Ash	50.5	48.3	47.5	41.7	1.08	1.53	0.004	0.014	0.237
Gross energy	63.1 ^{b}	66.7 ^{a}	56.2 ^{c}	57.2 ^{c}	0.41	0.58	<0.001	<0.001	0.037
Crude protein, CP	75.7 ^{ab}	76.8 ^{a}	74.5 ^{b}	72.1 ^{c}	0.49	0.70	<0.001	0.377	0.018
Total dietary fibre	**35.4** ^{c}	50.1 ^{a}	33.5 ^{c}	40.9 ^{b}	0.59	0.83	<0.001	<0.001	<0.001
Neutral detergent fibre	**28.1** ^{c}	40.2 ^{a}	28.0 ^{c}	31.6 ^{b}	0.69	0.97	<0.001	<0.001	<0.001
Soluble fibre	61.9	74.7	58.0	69.0	1.51	2.14	0.034	<0.001	0.637
Digestible energy, MJ/kg DM	11.38 ^{b}	11.78 ^{a}	10.64 ^{c}	10.46 ^{c}	0.075	0.106	<0.001	0.287	0.010
Digestible CP, g/kg DM	14.91 ^{a}	15.23 ^{a}	13.89 ^{b}	13.17 ^{c}	0.095	0.134	<0.001	0.147	<0.001
Ratio digestible [CP/energy], g/MJ	13.11	12.93	13.105	12.59	0.052	0.074	0.012	<0.001	0.069

[1] LIF-LSF = Low-insoluble fibre, Low-soluble fibre; LIF-HSF = Low-insoluble fibre, High-soluble fibre HIF-LSF = High-insoluble fibre, Low-soluble fibre; HIF-HSF = High-insoluble fibre, High-soluble fibre. $^{a–c}$ Fattening mean values in the same row with a different superscript differ $p < 0.050$.

Table 4. Effect of dietary level of insoluble and soluble fibre on nitrogen (N) balance from 28 to 62 days of age.

Item	Diets [1]				SEM		*p*-Value		
	Low-Insoluble Fibre		High-Insoluble Fibre						
	Low-Soluble Fibre LIF-LSF	High-Soluble Fibre LIF-HSF	Low-Soluble Fibre HIF-LSF	High-Soluble Fibre HIF-HSF	IF and SF	IF × SF	IF	SF	IF × SF
N	34	36	32	33					
kg $BW^{0.75}$	1.27	1.25	1.25	1.24	0.01	0.01	0.101	0.086	0.389
[2] Nitrogen intake, g/kg $BW^{0.75}$ d									
Ni,	2.98	2.97	3.23	3.10	0.03	0.04	<0.001	0.111	0.157
DNi	2.25 [b]	2.28 [b]	2.41 [a]	2.24 [b]	0.02	0.03	0.138	0.030	0.003
[3] Nitrogen retained, g/kg $BW^{0.75}$ d									
NR whole body	1.05	1.03	1.02	1.02	0.05	0.01	0.040	0.166	0.510
NR carcass	0.711	0.696	0.682	0.677	0.007	0.010	0.017	0.275	0.570
[4] Nitrogen efficiency									
NR whole body/Ni	0.353	0.351	0.318	0.330	0.004	0.006	<0.001	0.383	0.186
NR whole body/DNi	0.467 [a]	0.456 [a]	0.427 [b]	0.458 [a]	0.005	0.007	0.017	0.161	0.005
NR whole body/DNp	0.561 [a]	0.549 [a]	0.506 [b]	0.550 [a]	0.007	0.010	0.015	0.111	0.049
NR carcass/DNi	0.316 [a]	0.308 [a]	0.284 [b]	0.305 [a]	0.004	0.006	0.004	0.283	0.011
NR carcass/DNp	0.381 [a]	0.370 [a]	0.336 [b]	0.366 [a]	0.005	0.008	0.004	0.205	0.008
[5] Nitrogen losses, g/kg $BW^{0.75}$ d									
Skin and viscera	0.337	0.335	0.343	0.342	0.006	0.008	0.469	0.803	0.988
Faeces	0.725 [c]	0.691 [d]	0.824 [b]	0.865 [a]	0.008	0.011	<0.001	0.74	<0.001
Urine	1.21 [b]	1.25 [b]	1.38 [a]	1.22 [b]	0.02	0.03	0.057	0.082	0.002

[1] LIF-LSF = Low-insoluble fibre, Low-soluble fibre; LIF-HSF = Low-insoluble fibre, High-soluble fibre HIF-LSF = High-insoluble fibre, Low-soluble fibre; HIF-HSF = High-insoluble fibre, High-soluble fibre. [2] Ni (g Ni/kg $BW^{0.75}$ day): Nitrogen intake. DNi (g DNi/kg $BW^{0.75}$ day): Digestible N intake. [3] NR whole body (g/kg $BW^{0.75}$ day): retained N in the whole body. NR carcass (g/kg $BW^{0.75}$ day): retained N in carcass. [4] DNp (g DNp/kg $BW^{0.75}$ day): Intake of digestible N used for growth, obtained by the difference between DNi and the DN used for maintenance (0.464 g DNm/kg $BW^{0.75}$ day). Skin and viscera (g N/kg $BW^{0.75}$ day): (g N retained in the whole body − g retained in carcass)/kg $BW^{0.75}$ day. Faeces (g/kg $BW^{0.75}$ day): (Total N intake − DNi)/kg $BW^{0.75}$ day. [5] N lost as skin and viscera (g/kg $BW^{0.75}$ day) = (g NR in the whole body − g NR carcass)/kg $BW^{0.75}$ day. N excreted as faeces (g/kg $BW^{0.75}$ day) = (Ni − DNi)/kg $BW^{0.75}$ day. Urine (g/kg $BW^{0.75}$ day): (DNi − NR whole body)/kg $BW^{0.75}$ day. Weight at weaning (28 days) as a covariate. [a–d] Fattening mean values in the same row with a different superscript differ $p < 0.050$.

The DE intake was also the highest for the HIF-LSF group compared with the other three groups (1.15 vs. 1.10 MJ DE/kg $BW^{0.75}$ d; $p = 0.027$, Table 5). The energy retained in the whole body and in the carcass tended to decrease when NDF and SF increased ($p \leq 0.084$). Consequently, the LIF-LSF group had the highest DE retention efficiency in the whole body and in the carcass compared with the other three groups (0.330 vs. 0.302, and 0.204 vs. 0.185, $p \leq 0.029$). These differences were even higher for the retention efficiency of estimated DE used for growth in the whole body and in the carcass, decreasing with the increase of total dietary fibre, although showing the lowest values in rabbits fed the HIF-LSF diet ($p \leq 0.027$). Faecal energy losses increased with NDF and decreased with SF ($p < 0.001$), and were inversely proportional to the DE of the diets. The energetic losses as urine and heat production were minimal for the LIF-LSF group and maximal for the HIF-LSF group, showing intermediate values for the other two groups ($p = 0.014$).

The digestible mineral intake was the highest for the HIF-LSF group compared with the other three groups (4.59 vs. 4.01 g MIi/kg $BW^{0.75}$ day; $p < 0.001$, Table 6). The minerals retained in the whole body were not affected by treatments, but those retained in the carcass decreased by 2% when NDF increased (1.021 vs. 0.997 g MI/kg $BW^{0.75}$ day; $p < 0.001$). The LIF-LSF group showed the highest efficiency of retention of digestible minerals in the carcass, which decreased with both NDF and SF, although the HIF-LSF group showed the lowest value ($p < 0.001$). Faecal mineral losses increased with both the level of NDF and SF (by 11% and 8%, respectively; $p < 0.001$). Mineral urine losses evolved inversely to the carcass retention efficiency ($p < 0.001$).

The increase of NDF had no influence on the villous height in the LSF groups but tended to decrease it in LIF-HSF and increase it in the HIF-HSF groups ($p = 0.071$, Table 7). Treatments did not affect the crypt depth, the ratio villous height/crypt depth, and the number of goblet cells per villi in the jejunal mucosa. The increase of SF impaired the sucrose specific activity (by 21%; $p = 0.034$), while the increase of NDF tended to reduce it ($p = 0.059$).

Table 5. Effect of dietary level of insoluble and soluble fibre on energy (E) balance from 28 to 62 days of age.

Item	Diets [1]				SEM		*p*-Value		
	Low-Insoluble Fibre		High-Insoluble Fibre						
	Low-Soluble Fibre LIF-LSF	High-Soluble Fibre LIF-HSF	Low-Soluble Fibre HIF-LSF	High-Soluble Fibre HIF-HSF	IF and SF	IF × SF	IF	SF	IF × SF
N	34	36	32	33					
$BW^{0.75}$ d	1.27	1.25	1.25	1.24	0.01	0.01	0.101	0.086	0.389
[2] Energy intake, MJ/kg $BW^{0.75}$ d									
GEi	1.70	1.65	2.05	1.94	0.02	0.03	<0.001	0.002	0.24
DEi	1.08^{b}	1.10^{b}	1.15^{a}	1.11^{ab}	0.01	0.02	0.016	0.619	0.027
[3] Energy retained, kJ/kg $BW^{0.75}$ d									
ER whole body	354	339	338	333	4.04	5.73	0.068	0.067	0.405
ER carcass	219	209	206	203	2.83	4.01	0.030	0.084	0.348
[4] Energy efficiency [1]									
ER whole body/GEi	0.208	0.207	0.165	0.173	0.003	0.004	<0.001	0.415	0.247
ER whole body/DEi	0.330^{a}	0.311^{b}	0.294^{b}	0.302^{b}	0.004	0.006	<0.001	0.350	0.029
ER whole body/DEp	0.677^{a}	0.63[illegible] ^{ab}	0.557^{c}	0.598^{bc}	0.014	0.020	<0.001	0.896	0.027
ER carcass/DEi	0.204^{a}	0.191^{b}	0.179^{b}	0.184^{b}	0.003	0.004	<0.001	0.323	0.024
ER carcass/DEp	0.419^{a}	0.387^{a}	0.339^{b}	0.365^{ab}	0.009	0.012	<0.001	0.831	0.017
[5] Energy losses, MJ/kg $BW^{0.75}$ d									
Skin and viscera,	0.135	0.132	0.131	0.130	0.003	0.004	0.647	0.331	0.794
Faeces	0.628	0.551	0.896	0.828	0.007	0.010	<0.001	<0.001	0.624
Urine + heat production	0.721^{c}	0.765^{bc}	0.814^{a}	0.777^{ab}	0.012	0.017	0.004	0.837	0.014

[1] LIF-LSF = Low-insoluble fibre, Low-soluble fibre; LIF-HSF = Low-insoluble fibre, High-soluble fibre HIF-LSF = High-insoluble fibre, Low-soluble fibre; HIF-HSF = High-insoluble fibre, High-soluble fibre. [2] GEi: Gross Energy intake (MJ/kg $BW^{0.75}$ day). DEi: Digestible Energy intake (MJ/kg $BW^{0.75}$ day). [3] ER whole body (kJ/kg $BW^{0.75}$ day): GE retained in the whole body. ER carcass (kJ/kg $BW^{0.75}$ day): GE retained in carcass. [4] DEp (MJ/kg $BW^{0.75}$ day): Intake of DE used for growth, obtained by the difference between DEi and the DE used for maintenance (430 kJ DE/kg $BW^{0.75}$ day). [5] Skin and viscera (MJ/kg $BW^{0.75}$ day): (MJ GE retained in the whole body − MJ GE retained in carcass)/kg $BW^{0.75}$ day. Faeces (MJ/kg $BW^{0.75}$ and day): (GEi − DEi)/kg $BW^{0.75}$ day. Urine + heat production (MJ/kg $BW^{0.75}$ d): (DEi − GE retained in carcass − GE lost in skin and viscera)/kg $BW^{0.75}$ day. Weight at weaning (28 days) as a covariate. [a–c] Fattening mean values in the same column with a different superscript differ $p < 0.050$.

Table 6. Effect of dietary level of insoluble and soluble fibre on mineral (MI) balance from 28 to 62 days of age.

Item	Diets [1]				SEM		*p*-Value		
	Low-Insoluble Fibre		High-Insoluble Fibre						
	Low-Soluble Fibre LIF-LSF	High-Soluble Fibre LIF-HSF	Low-Soluble Fibre HIF-LSF	High-Soluble Fibre HIF-HSF	IF and SF	IF × SF	IF	SF	IF × SF
N	34	36	32	33					
[2] Mineral intake, g/kg $BW^{0.75}$ d									
MIi,	7.92	8.34	9.66	9.66	0.09	0.13	<0.001	0.097	0.095
DMIi	3.97 ^{b}	4.04 ^{b}	4.59 ^{a}	4.02 ^{b}	0.04	0.06	<0.001	<0.001	<0.001
[3] Mineral retained, g/kg $BW^{0.75}$ d									
MIR whole body	1.25	1.22	1.23	1.23	0.01	0.01	0.337	0.223	0.068
MIR carcass	1.027	1.016	0.998	0.996	0.005	0.007	<0.001	0.315	0.531
Minerals efficiency									
MIR whole body/MIi	0.159 ^{a}	0.148 ^{b}	0.127 ^{c}	0.128 ^{c}	0.002	0.002	<0.001	0.032	0.009
MIR whole body/DMIi	0.318 ^{a}	0.306 ^{a}	0.268 ^{b}	0.308 ^{a}	0.003	0.005	<0.001	0.005	<0.001
MIR in carcass/DMIi	0.261 ^{a}	0.254 ^{ab}	0.218 ^{c}	0.249 ^{b}	0.003	0.004	<0.001	0.036	<0.001
[4] Mineral losses, g/kg $BW^{0.75}$ d									
Skin and viscera,	0.227	0.207	0.227	0.236	0.007	0.011	0.192	0.594	0.160
Faeces	3.95	4.30	5.07	5.63	0.05	0.07	<0.001	<0.001	0.110
Urine	2.71 ^{b}	2.81 ^{b}	3.36 ^{a}	2.79 ^{b}	0.04	0.06	<0.001	<0.001	<0.001

[1] LIF-LSF = Low-insoluble fibre, Low-soluble fibre; LIF-HSF = Low-insoluble fibre, High-soluble fibre HIF-LSF = High-insoluble fibre, Low-soluble fibre; HIF-HSF = High-insoluble fibre, High-soluble fibre. [2] MIi (g MIi/kg $BW^{0.75}$ day): Mineral intake. DMIi (g DMIi/kg $BW^{0.75}$ day): Digestible mineral intake. [3] MIR whole body (g/kg $BW^{0.75}$ day): Mineral retained in the whole body. MIR carcass (g/kg $BW^{0.75}$ day): Mineral retained in carcass. [4] Skin and viscera (g Min/kg $BW^{0.75}$ day): (g MIR in the whole body − g MIR in carcass)/kg $BW^{0.75}$ day. Faeces (g/kg $BW^{0.75}$ day): (MIi − DMIi)/kg $BW^{0.75}$ d. Faeces (g/kg $BW^{0.75}$ day) = (MINi − DMIi)/kg $BW^{0.75}$ day. Urine (g/kg $BW^{0.75}$ day): (DMIi − MIR whole body)/kg $BW^{0.75}$ day. Weight at weaning (28 day) as a covariate. [a–c] Fattening mean values in the same row with a different superscript differ $p < 0.050$.

Table 7. Effect of dietary level of insoluble and soluble fibre on mucosa histology and enzyme activity of 39-day-old rabbits.

Item	Diets [1]				SEM		p-Value		
	Low-Insoluble Fibre		High Insoluble Fibre						
	Low-Soluble Fibre LIF-LSF	High-Soluble Fibre LIF-HSF	Low-Soluble Fibre HIF-LSF	High-Soluble Fibre HIF-HSF	IF and SF	IF × SF	IF	SF	IF × SF
N	10	7	9	7					
Villous height, µm	380	340	383	422	14.8	20.8	0.052	0.995	0.071
Crypt depth, µm	144	156	147	160	5.45	7.69	0.667	0.115	0.989
Ratio villous height/crypt depth	2.71	2.20	2.67	2.70	0.15	0.22	0.293	0.290	0.232
Goblet cells, n°/villous	11.1	9.86	10.3	11.8	1.06	1.49	0.707	0.923	0.350
Mucose protein, mg/g of tissue	61.5	46.2	57.6	52.2	4.81	6.80	0.878	0.136	0.470
Sucrose activity, µmol of glucose/mg of protein	273	212	218	174	17.0	24.0	0.059	0.034	0.703

[1] LIF-LSF = Low-insoluble fibre, Low-soluble fibre; LIF-HSF = Low-insoluble fibre, High-soluble fibre HIF-LSF = High-insoluble fibre, Low-soluble fibre; HIF-HSF = High-insoluble fibre, High-soluble fibre.

4. Discussion

Digestive disorders are one of the main causes of mortality in growing rabbits, with antibiotic treatment being the usual way to control them [26]. One of the nutritional strategies to limit ERE is the inclusion of moderate levels of SF [3,6,27,28], while the increase of NDF might increase its incidence [15,16]. However, in this study, a very low mortality rate was reported, avoiding any conclusion about treatments, and suggesting a stronger influence of other environmental factors than the diet. Even the higher dietary protein levels than those expected did not trigger an ERE outbreak, although a high protein level is recognized as an important risk factor [27,29–31]. Under these good sanitary conditions, there was no positive effect of the level of SF on the mucosa morphology (ratio villous height/crypth depth and number of goblet cells) or functionality (sucrose-specific activity), which contrast with the positive effect reported when there is an ERE outbreak [1,5]. The trend observed for the interaction in the villous height suggests a negative influence of SF in the LIF groups but positive in the HIF groups, an observation that will require further confirmation. The negative influence of NDF on the sucrose-specific activity, similarly to that of SF, might be partially associated to the reduction of starch intake, as they were positively correlated ($n = 4$; $r = 0.90$; $p = 0.099$), rather than to mucosa damage.

The fibre digestibility depends on microbial activity, fermentation time, and digestion rate, which are all affected by the chemical composition of dietary fibre (mainly by the degree of lignification of NDF and the SF content) and particle size [32]. The digestibility of NDF of the LIF-LSF diet was low and close to other low insoluble-soluble fibre diets with medium-high DE concentrations [3,8,27,28,33]. The increase of NDF when SF was low (HIF-LSF vs. LIF-LSF) did not modify NDF digestibility, although the degree of lignification and feed intake increased, indicating that it probably accelerated the rate of passage [34,35]. This lack of effect might be associated to the fast digestion rate of the degradable fraction of NDF, which can be degraded/solubilized up to 50% before the caecum [9]. Comparatively, it was surprising that soluble fibre digestibility decreased from the LIF-LSF to the HIF-LSF group, which might account for the potential reduction in the mean retention time and the changes in the ingredient proportions. However, a higher incidence of these factors on the NDF than on the SF fraction was expected. The increase of SF when NDF is low (LIF-HSF vs. LIF-LSF) improved both NDF and SF digestibility in agreement with previous studies [7,9,23]. It is explained by the faster digestion rate of SF, similar to that observed in vitro for SBP when compared with straw [36]. The increase of NDF when SF is high (HIF-HSF vs. LIF-HSF) improved also the NDF and SF digestibility as expected [37,38] but the NDF to a lesser extent. It might be linked to the faster rate of passage associated to the HIF groups, which might limit the fermentation of SBP fibre and accordingly its nutritive value. This effect was already observed [12], when a higher DE value for SBP using a high energy basal diet than a low one was reported.

The increase of NDF led to a higher feed intake, and impaired feed efficiency as expected. On the opposite, the increase of SF did not modify feed efficiency, although it tended to reduce feed intake and the growth rate. The replacement of starch mainly by SF and the absence of ERE might be behind this lack of effect on feed efficiency as observed previously [26], although other authors found a positive effect [8,37,39]. These contradictory results are not explained by differences in the acid detergent fibre content [7,40], and do not seem associated to the health status. However, a potential improvement of the nitrogenous retention efficiency, and a reduction of nitrogen losses might be behind the positive effect of SF on feed efficiency [8]. In contrast, the inclusion of SF in the substitution of NDF, and minor proportions of starch and fat, usually improves the feed efficiency [3,38,41]. The magnitude of the feed intake reduction with the inclusion of SF widely differs among studies, with no great influence on DEi in some of them, but in others, where the level of SBP inclusion exceeded 15–30%, compromised DEi (and even DNi), impairing the feed conversion ratio and dressing out percentage [6,8,10,42,43]. It was related to the accumulation of digesta in the caecum derived from a combination of factors like the level and type of NDF, and effective particle size [44].

In this study, rabbits tended to grow slower either with the increase of NDF or SF or both, even when they had a similar or even higher DE and DN intake. These results agree with the trend to reduce the whole-body protein and fat accretion with the increase of total dietary fibre, with the reduction being more pronounced for fat than for protein. It led to a progressive reduction of the nitrogen and energy retained in the whole body and in the carcass, even when diets were formulated to meet all nutrient requirements, which contrasts with the lack of an effect of SF on these traits observed in a previous study [8]. These results are explained by the trend to impair the retention efficiency of the DE used for growth (DEp) both in the whole body and in the carcass when total dietary fibre increased. The reduction of the energetic efficiency is parallel to the increase of energetic losses as urine and heat production from the LIF-LSF to the HIF-HSF group (excluding the HIF-LSF group), most probably related to the heat increment associated to the use of nutrients for growth and that associated with the increase of the fermentative activity, as the N losses in urine were similar among these groups. The increase of total dietary fibre (from the LIF-LSF to the HIF-HSF group) implied a higher intake of digestible total dietary fibre (by 65%) and a lower intake of starch (by 55%), accompanied in the HIF groups by an increase of fat intake (by 140%). The higher metabolic efficiency of glucose compared to volatile fatty acids, and the potential need in the HIF groups to obtain glucose from amino acids seemed to be behind the energetic efficiency impairment observed with the increase of total dietary fibre. In relation to fat, other authors found no influence of the increase of fat intake on the overall DE or DE efficiency for growth [45], suggesting a minor influence on the energetic efficiency results of this study. The impairment of the efficiency of the retention of DE (used for growth) was not related with the efficiency of retention of DN (used for growth), which was similar among groups (excluding the HIF-LSF group, which will deserve a separate comment). These results contrast with the lack of an effect of SF on the retained energy and nitrogen observed previously [8], which might be explained by the higher starch and lower fermentable fibre intake obtained in their study [8,23]. Besides, the same authors observed a positive effect of the increase of SF on nitrogen metabolism due to the reduction of urinary nitrogen losses, probably derived from a shift of urinary nitrogen excretion to faecal excretion. This effect was not observed in this study probably due to the excess of dietary digestible protein, despite the increase of the fermentable fibre intake with the level of total dietary fibre. Rabbits fed the HIF-LSF diet showed the worst efficiencies of DN and of DE used for growth, which were lower than those expected according to its NDF and SF levels. It is mainly explained by the high DN intake, derived from the combination of the highest feed intake and a high dietary DN/DE ratio, that led to the highest urinary nitrogen (and energetic) losses. The latter are associated with higher heat production derived from urea synthesis, which helps to explain the trend to increase energetic losses as urine and heat production in rabbits fed HIF-LSF compared with those of the HIF-HSF group ($p = 0.112$). In fact, when experiments with wide differences in dietary digestible protein content are combined, a close relationship between the DN intake and urinary nitrogen losses is found (Figure 1).

The influence of dietary fibre on mineral balance in rabbits has been scarcely studied. The mineral retention in the carcass was impaired with the increase of NDF, although the digestible mineral intake was similar (HIF-HSF) or even higher (HIF-LSF group). It led to a reduction on the efficiency of retention of digested minerals with the increase of total dietary fibre (except for the HIF-LSF group). This result agreed with the reduction of the serum mineral concentrations (per unit of mineral ingested) in pregnant sows when dietary fibre increased [46]. However, the increase of dietary fibre did not influence all minerals in the same way. In fact, the apparent absorption of calcium, phosphorous, and magnesium was not influenced by the fibre level, while it impaired that of sodium and potassium in pigs [47]. In rabbits, the type of fibre (alfalfa, olive pulp, grape pulp) also influenced the apparent absorption of different minerals, but no effect was reported on the plasma levels of most elements [48].

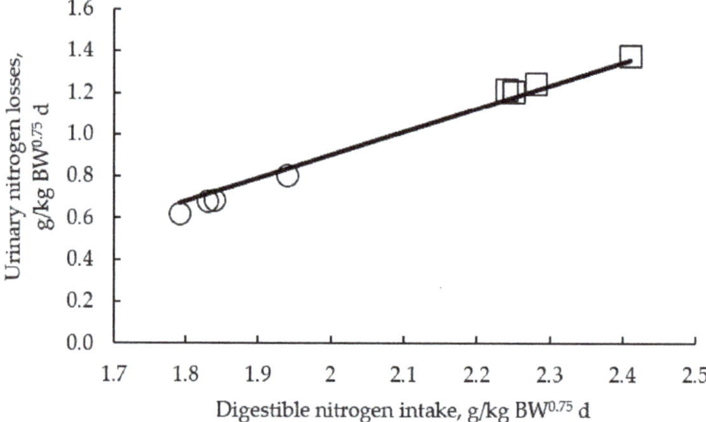

Figure 1. Relationship between digestible nitrogen intake and urinary nitrogen losses during fattening in rabbits (Urinary nitrogen losses = −1.31(±0.13) + 1.11(±0.061) digestible nitrogen intake; N = 8; $p < 0.001$. Original observations plotted with the mean regression line across studies: ○ Delgado et al., 2018. □ Current study).

5. Conclusions

The increase of either NDF or SF in diets for healthy rabbits had no positive effect in terms of growth traits, retention efficiency of digestible nitrogen energy and minerals, and mucosa morphology, once the NDF requirements for growing rabbits are met. Further research is required to identify what is the optimal level of NDF to combine with SF under an enteropathy outbreak.

Supplementary Materials: The following are available online at http://www.mdpi.com/2076-2615/10/8/1346/s1: Table S1: Effect of dietary level of insoluble and soluble fibre on the whole body chemical composition and energy content of weaning rabbits at 28 and 62 days of age. Table S2: Effect of dietary level of insoluble and soluble fibre on carcass chemical composition and energy content of weaning rabbits at 28 and 62 days of age. Table S3: Effect of dietary level of insoluble and soluble fibre on daily gain of whole body components and their proportions in the daily gain from 28 to 62 days of age. Table S4: Effect of dietary level of insoluble fibre and soluble fibre on daily gain of carcass components and their proportions in the carcass daily gain from 28 to 62 days.

Author Contributions: J.G., R.C. and N.N. contributed to concept and design of the experiment. C.F.-K., N.N., R.D., C.O.-V., T.N., R.A.-S.A. and J.G. performed the animal trial including the faecal and histological sampling. C.F.-K. conducted the chemical and histological determinations (helped by R.D., C.O.-V., T.N. and R.A.-S.A.), statistical analysis, and prepared the tables. All the authors contributed to the interpretation of the results. C.F.-K. and J.G. wrote the manuscript with the collaboration of N.N. and R.C. All authors have read and agreed to the published version of the manuscript.

Funding: This research was financed by the project MINECO-FEDER (AGL2015-66485-R and the pre-doctoral contract BES-2016-076649 obtained by Carlos Farías-Kovac) awarded by the Ministerio de Economía y Competitividad (Spain).

Conflicts of Interest: The authors declare no conflict of interest. The funders had no role in the design of the study; in the collection, analyses, or interpretation of data; in the writing of the manuscript, or in the decision to publish the results.

References

1. Gómez-Conde, M.S.; García, J.; Chamorro, S.; Eiras, P.; Rebollar, P.G.; Pérez de Rozas, A.; Badiola, I.; de Blas, C.; Carabaño, R. Neutral detergent-soluble fiber improves gut barrier function in twenty-five-day-old weaned rabbits1. *J. Anim. Sci.* **2007**, *85*, 3313–3321. [CrossRef] [PubMed]

2. El abed, N.; Delgado, R.; Abad-Guamán, R.; Romero, C.; Villamide, M.J.; Menoyo, D.; Carabaño, R.; García, J. Soluble and insoluble fibre from sugar beet pulp enhance intestinal mucosa morphology in young rabbits. In *Book of Abstracts of the 62nd Annual Meeting of the European Association for Animal Production*; EAAP Book of Abstracts; Wageningen Academic Publishers: Stavanger, Norway, 2011; Volume 17, p. 159, ISBN 978-90-8686-731-8.
3. Gómez-Conde, M.S.; de Rozas, A.P.; Badiola, I.; Pérez-Alba, L.; de Blas, C.; Carabaño, R.; García, J. Effect of neutral detergent soluble fibre on digestion, intestinal microbiota and performance in twenty five day old weaned rabbits. *Livest. Sci.* **2009**, *125*, 192–198. [CrossRef]
4. Rodríguez-Romero, N.; Abecia, L.; Fondevila, M. Bacterial profile from caecal contents and soft faeces in growing rabbits given diets differing in soluble and insoluble fibre levels. *Anaerobe* **2012**, *18*, 602–607. [CrossRef]
5. El abed, N.; Badiola, I.; Pérez de Rozas, A.; González, J.; Menoyo, D.; Carabaño, R.; García, J. Effect of soluble and insoluble fractions of sugar beet pulp on ileal and caecal microbiota of rabbits after weaning. *World Rabbit Sci.* **2013**, *21*, 207–208. [CrossRef]
6. Martínez-Vallespín, B.; Martínez-Paredes, E.; Ródenas, L.; Cervera, C.; Pascual, J.J.; Blas, E. Combined feeding of rabbit female and young: Partial replacement of starch with acid detergent fibre or/and neutral detergent soluble fibre at two protein levels. *Livest. Sci.* **2011**, *141*, 155–165. [CrossRef]
7. Trocino, A.; García Alonso, J.; Carabaño, R.; Xiccato, G. A meta-analysis on the role of soluble fibre in diets for growing rabbits. *World Rabbit Sci.* **2013**, *21*. [CrossRef]
8. Delgado, R.; Nicodemus, N.; Abad-Guamán, R.; Sastre, J.; Menoyo, D.; Carabaño, R.; García, J. Effect of dietary soluble fibre and n-6/n-3 fatty acid ratio on growth performance and nitrogen and energy retention efficiency in growing rabbits. *Anim. Feed Sci. Technol.* **2018**, *239*, 44–54. [CrossRef]
9. Abad-Guamán, R. Identification of the Method to Quantify Soluble Fibre and the Effect of the Source of Fibre on the Ileal and Faecal Digestibility of Soluble and Insoluble Fibre in Rabbits. Ph.D. Thesis, E.T.S.I. Agrónomos (Universidad Politécnica de Madrid), Madrid, Spain, 2015.
10. Garcia, G.; Galvez, J.F.; de Blas, J.C. Effect of substitution of sugarbeet pulp for barley in diets for finishing rabbits on growth performance and on energy and nitrogen efficiency. *J. Anim. Sci.* **1993**, *71*, 1823–1830. [CrossRef]
11. Carabaño, R.; Motta-Ferreira, W.; de Blas, J.C.; Fraga, M.J. Substitution of sugarbeet pulp for alfalfa hay in diets for growing rabbits. *Anim. Feed Sci. Technol.* **1997**, *65*, 249–256. [CrossRef]
12. de Blas, C.; Villamide, M.J. Nutritive value of beet and citrus pulps for rabbits. *Anim. Feed Sci. Technol.* **1990**, *31*, 239–246. [CrossRef]
13. García, J., Carabaño, R.; de Blas, J.C. Effect of fiber source on cell wall digestibility and rate of passage in rabbits. *J. Anim. Sci.* **1999**, *77*, 898. [CrossRef] [PubMed]
14. Gidenne, T. Effets d'une réduction de la teneur en fibres alimentaires sur le transit digestif du lapin. Comparaison et validation de modèles d'ajustement des cinétiques d'excrétion fécale des marqueurs. *Reprod. Nutr. Dev.* **1994**, *34*, 295–306. [CrossRef] [PubMed]
15. Gutiérrez, I.; Espinosa, A.; García, J.; Carabaño, R.; De Blas, J.C. Effect of levels of starch, fiber, and lactose on digestion and growth performance of early-weaned rabbits. *J. Anim. Sci.* **2002**, *80*, 1029–1037. [CrossRef] [PubMed]
16. Tazzoli, M.; Trocino, A.; Birolo, M.; Radaelli, G.; Xiccato, G. Optimizing feed efficiency and nitrogen excretion in growing rabbits by increasing dietary energy with high-starch, high-soluble fibre, low-insoluble fibre supply at low protein levels. *Livest. Sci.* **2015**, *172*, 59–68. [CrossRef]
17. BOE.es—Documento BOE-A-2013-1337. Available online: https://www.boe.es/diario_boe/txt.php?id=BOE-A-2013-1337 (accessed on 21 May 2020).
18. de Blas, C.; Mateos, G.G. Feed formulation. In *Nutrition of the Rabbit*, 3rd ed.; CABI Publishing CAB International: Wallingford, UK, 2020; pp. 243–253, ISBN 978-1-78924-127-3.
19. Saiz, A.; García-Ruiz, A.I.; Fernández, A.; Nicodemus, N. Application of Bioelectrical impedance analysis (BIA) to study the carcass chemical composition of rabbits from 35 to 63 d of age. *World Rabbit Sci.* **2013**, *21*, 208–209. [CrossRef]
20. Saiz, A.; García-Ruiz, A.I.; Fuentes-Pila, J.; Nicodemus, N. Application of bioelectrical impedance analysis to assess rabbit's body composition from 25 to 77 days of age1. *J. Anim. Sci.* **2017**, *95*, 2782–2793. [CrossRef]

21. Pérez, J.M.; Cervera, C.; Falcao E Cunha, L.; Maertens, L.; Villamide, M.J.; Xiccato, G. European ring-test on in vivo determination of digestibility in rabbits: Reproducibility of a reference method in comparison with domestic laboratory procedures. *World Rabbit Sci.* **1995**, *3*, 171–178. [CrossRef]
22. Xiccato, G.; Trocino, A. Energy and protein metabolism and requirements. In *Nutrition of the Rabbit*, 3rd ed.; CABI Publishing CAB International: Wallingford, UK, 2020; pp. 89–125, ISBN 978-1-78924-127-3.
23. Delgado, R.; Menoyo, D.; Abad-Guamán, R.; Nicodemus, N.; Carabaño, R.; García, J. Effect of dietary soluble fibre level and n-6/n-3 fatty acid ratio on digestion and health in growing rabbits. *Anim. Feed Sci. Technol.* **2019**, *255*, 114222. [CrossRef]
24. Horwitz, W.; AOAC International (Eds.) *Official Methods of Analysis of AOAC International*, 18th ed.; Current Through rev. 1, 2006; AOAC International: Gaithersburg, MD, USA, 2006; ISBN 978-0-935584-77-6.
25. Mertens, D.R. Gravimetric determination of amylase-treated neutral detergent fiber in feeds with refluxing in beakers or crucibles: Collaborative study. *J. AOAC Int.* **2002**, *85*, 1217–1240.
26. Solans, L.; Arnal, L.J.; Sanz, C.; Benito, A.; Chacón, G.; Alzuguren, O.; Fernández, B.A. Rabbit Enteropathies on commercial farms in the iberian peninsula: Etiological agents identified in 2018–2019. *Animals* **2019**, *9*, 1142. [CrossRef]
27. Xiccato, G.; Trocino, A.; Majolini, D.; Fragkiadakis, M.; Tazzoli, M. Effect of decreasing dietary protein level and replacing starch with soluble fibre on digestive physiology and performance of growing rabbits. *Animal* **2011**, *5*, 1179–1187. [CrossRef] [PubMed]
28. Grueso, I.; De Blas, J.C.; Cachaldora, P.; Mendez, J.; Losada, B.; García-Rebollar, P. Combined effects of supplementation of diets with hops and of a substitution of starch with soluble fiber on feed efficiency and prevention of digestive disorders in rabbits. *Anim. Feed Sci. Technol.* **2013**, *180*, 92–100. [CrossRef]
29. Chamorro, S.; Gómez-Conde, M.S.; Pérez de Rozas, A.M.; Badiola, I.; Carabaño, R.; De Blas, J.C. Effect on digestion and performance of dietary protein content and of increased substitution of lucerne hay with soya-bean protein concentrate in starter diets for young rabbits. *Animal* **2007**, *1*, 651. [CrossRef] [PubMed]
30. Gidenne, T.; Kerdiles, V.; Jehl, N.; Arveux, P.; Eckenfelder, B.; Briens, C.; Stephan, S.; Fortune, H.; Montessuy, S.; Muraz, G. Protein replacement by digestible fibre in the diet of growing rabbits: 2-Impact on performances, digestive health and nitrogen output. *Anim. Feed Sci. Technol.* **2013**, *183*, 142–150. [CrossRef]
31. Gutiérrez, I.; Espinosa, A.; García, J.; Carabaño, R.; De Blas, C. Effect of protein source on digestion and growth performance of early-weaned rabbits. *Anim. Res.* **2003**, *52*, 461–471. [CrossRef]
32. Gidenne, T.; Carabaño, R.; Abad-Guamán, R.; García, J.; de Blas, C. Fibre digestion. In *Nutrition of the Rabbit*, 3rd ed.; CABI Publishing CAB International: Wallingford, UK, 2020; pp. 69–88, ISBN 978-1-78924-127-3.
33. Gidenne, T.; Perez, J.-M. Replacement of digestible fibre by starch in the dietof the growing rabbit. I. Effects on digestion, rate of passage and retention of nutrients. *Ann. Zootech.* **2000**, *49*, 357–368. [CrossRef]
34. Nicodemus, N.; Carabaño, R.; García, J.; Méndez, J.; de Blas, C. Performance response of lactating and growing rabbits to dietary lignin content. *Anim. Feed Sci. Technol.* **1999**, *80*, 43–54. [CrossRef]
35. Gidenne, T.; Arveux, P.; Madec, O. The effect of the quality of dietary lignocellulose on digestion, zootechnical performance and health of the growing rabbit. *Anim. Sci.* **2001**, *73*, 97–104. [CrossRef]
36. Ocasio-Vega, C.; Abad-Guamán, R.; Delgado, R.; Carabaño, R.; Carro, M.D.; García, J. Effect of cellobiose supplementation and dietary soluble fibre content on *in vitro* caecal fermentation of carbohydrate-rich substrates in rabbits. *Arch. Anim. Nutr.* **2018**, *72*, 221–238. [CrossRef]
37. Caîsin, L.; Martínez-Paredes, E.; Ródenas, L.; Moya, V.J.; Pascual, J.J.; Cervera, C.; Blas, E.; Pascual, M. Effect of increasing lignin in isoenergetic diets at two soluble fibre levels on digestion, performance and carcass quality of growing rabbits. *Anim. Feed Sci. Technol.* **2020**, *262*, 114396. [CrossRef]
38. Trocino, A.; Fragkiadakis, M.; Majolini, D.; Carabaño, R.; Xiccato, G. Effect of the increase of dietary starch and soluble fibre on digestive efficiency and growth performance of meat rabbits. *Anim. Feed Sci. Technol.* **2011**, *165*, 265–277. [CrossRef]
39. Trocino, A.; Fragkiadakis, M.; Majolini, D.; Tazzoli, M.; Radaelli, G.; Xiccato, G. Soluble fibre, starch and protein level in diets for growing rabbits: Effects on digestive efficiency and productive traits. *Anim. Feed Sci. Technol.* **2013**, *180*, 73–82. [CrossRef]
40. Villamide, M.J.; Carabaño, R.; Maertens, L.; Pascual, J.; Gidenne, T.; Falcao-E-Cunha, L.; Xiccato, G. Prediction of the nutritional value of European compound feeds for rabbits by chemical components and in vitro analysis. *Anim. Feed Sci. Technol.* **2009**, *150*, 283–294. [CrossRef]

41. Trocino, A.; Fragkiadakis, M.; Radaelli, G.; Xiccato, G. Effect of dietary soluble fibre level and protein source on growth, digestion, caecal activity and health of fattening rabbits. *World Rabbit Sci.* **2010**, *18*, 199–210. [CrossRef]
42. Garcia, G.; Galvez, J.F.; De Blas, J.C. Substitution of barley grain by sugar-beet pulp in diets for finishing rabbits. 1. Effect on energy and nitrogen balance. *J. Appl. Rabbit Res.* **1992**, *15*, 1008.
43. Garcia, G.; Galvez, J.F.; De Blas, J.C. Substitution of barley grain by sugar-beet pulp in diets for finishing rabbits. 2. Effect on growth performance. *J. Appl. Rabbit Res.* **1992**, *15*, 1017.
44. García, J.; Gidenne, T.; Gidenne, T.; Falcao-e-Cunha, L.; de Blas, C. Identification of the main factors that influence caecal fermentation traits in growing rabbits. *Anim. Res.* **2002**, *51*, 165–173. [CrossRef]
45. Fernández, C.; Fraga, M.J. Effect of fat inclusion in diets for rabbits on the efficiency of digestible energy and protein utilization. *World Rabbit Sci.* **2010**, *4*. [CrossRef]
46. Girard, C.L.; Robert, S.; Matte, J.J.; Farmer, C.; Martineau, G.P. Influence of high fibre diets given to gestating sows on serum concentrations of micronutrients. *Livest. Prod. Sci.* **1995**, *43*, 15–26. [CrossRef]
47. Stanogias, G.; Pearce, G.R.; Alifakiotis, T.; Michaelidis, J.; Pappa-Michaelidis, B. Effects of dietary concentration and source of fibre on the apparent absorption of minerals by pigs. *Anim. Feed Sci. Technol.* **1994**, *47*, 287–295. [CrossRef]
48. Tortuero, F.; Rioperez, J.; Cosin, C.; Barrera, J.; Rodriguez, M.L. Effects of dietary fiber sources on volatile fatty acid production, intestinal microflora and mineral balance in rabbits. *Anim. Feed Sci. Technol.* **1994**, *48*, 1–14. [CrossRef]

Article

Antimicrobial Effects of Black Soldier Fly and Yellow Mealworm Fats and Their Impact on Gut Microbiota of Growing Rabbits

Sihem Dabbou [1,2], Ilario Ferrocino [3], Laura Gasco [3,*], Achille Schiavone [4], Angela Trocino [5], Gerolamo Xiccato [6], Ana C. Barroeta [7], Sandra Maione [4], Dominga Soglia [4], Ilaria Biasato [3], Luca Cocolin [3], Francesco Gai [8] and Daniele Michele Nucera [3]

1. Center Agriculture Food Environment (C3A), University of Trento, Via E. Mach 1, 38010 San Michele all'Adige, Italy; sihem.dabbou@unitn.it
2. Research and Innovation Centre, Fondazione Edmund Mach, 38010 San Michele all'Adige, Italy
3. Department of Agricultural, Forest and Food Sciences, University of Torino, Largo Paolo Braccini 2, 10095 Grugliasco, Italy; ilario.ferrocino@unito.it (I.F.); ilaria.biasato@unito.it (I.B.); lucasimone.cocolin@unito.it (L.C.); daniele.nucera@unito.it (D.M.N.)
4. Department of Veterinary Science, University of Turin, Largo P. Braccini 2, 10095 Grugliasco, Italy; achille.schiavone@unito.it (A.S.); sandra.maione@unito.it (S.M.); dominga.soglia@unito.it (D.S.)
5. Department of Comparative Biomedicine and Food Science, University of Padova, Viale dell'Università 16, Legnaro, 35020 Padova, Italy; angela.trocino@unipd.it
6. Department of Agronomy, Food, Natural Resources, Animal, and Environment, University of Padova, Viale dell'Università 16, 35020 Legnaro, Padova, Italy; gerolamo.xiccato@unipd.it
7. Nutrition and Animal Welfare Service, Department of Animal and Food Science, Faculty of Veterinary Medicine, Autonomous University of Barcelona, Bellaterra, 08193 Barcelona, Spain; Ana.Barroeta@uab.cat
8. Institute of Sciences of Food Production, National Research Council, Largo Paolo Braccini 2, 10095 Grugliasco, Italy; francesco.gai@ispa.cnr.it

* Correspondence: laura.gasco@unito.it

Received: 26 June 2020; Accepted: 24 July 2020; Published: 28 July 2020

Simple Summary: The use of insect lipids as an alternative ingredient is an emergent topic in animal nutrition due to their antimicrobial activities. The present study evaluated the in vitro antimicrobial activities of two insect fats (black soldier fly (*Hermetia illucens* (HI) fat and yellow mealworm (*Tenebrio molitor* (TM) fat) and their effect as a total substitute for dietary soybean oil in cecal fermentation and gut microbiota of growing rabbits. The obtained results showed the potential of HI and TM fats as an antibacterial feed ingredient with a positive influence on the rabbit cecal microbiota. HI and TM fats therefore may be a sustainable lipid alternative to soybean oil in rabbit nutrition with possible interesting applications in the feed industry.

Abstract: This study aimed to evaluate the in vitro antimicrobial activities of two types of insect fats extracted from black soldier fly larvae (HI, *Hermetia illucens* L.) and yellow mealworm larvae (TM, *Tenebrio molitor* L.) and their effects as dietary replacement of soybean oil (S) on cecal fermentation pattern, and fecal and cecal microbiota in rabbits. A total of 120 weaned rabbits were randomly allotted to three dietary treatments (40 rabbits/group) —a control diet (C diet) containing 1.5% of S and two experimental diets (HI diet (HID) and TM diet (TMD)), where S was totally substituted by HI or TM fats during the whole trial that lasted 41 days. Regarding the in vitro antimicrobial activities, HI and TM fats did not show any effects on *Salmonella* growth. *Yersinia enterocolitica* showed significantly lower growth when challenged with HI fats than the controls. The insect fat supplementation in rabbit diets increased the contents of the cecal volatile fatty acids when compared to the control group. A metataxonomic approach was adopted to investigate the shift in the microbial composition as a function of the dietary insect fat supplementation. The microbiota did not show a clear separation as a function of the inclusion, even if a specific microbial signature was observed. Indeed, HI and TM fat supplementation enriched the presence of *Akkermansia* that was found to be correlated with

NH3-N concentration. An increase in *Ruminococcus*, which can improve the immune response of the host, was also observed. This study confirms the potential of HI and TM fats as antibacterial feed ingredients with a positive influence on the rabbit cecal microbiota, thus supporting the possibility of including HI and TM fats in rabbit diets.

Keywords: insect fat; *Hermetia illucens*; *Tenebrio molitor*; gut microbiota; antimicrobial effect; rabbit feeding

---

1. Introduction

In rabbit production, a high mortality, which can reach 80%, due to gastrointestinal disorders and epizootic enteropathy is the major health concern [1]. The rabbit's digestive health and physiology, as well as the immune system, are based on its abundant cecal microbiota [2,3]. The intestinal microbial community is abundant, since it contains about 100 to 1000 billion microorganisms per gram of digesta. Its diversity and complexity is very high, with about a thousand different species. Bacteria predominate with a 10^{11} to 10^{12} bacteria/g cecal content [4]. Cecal microbiota of rabbit species is dominated by phylum Firmicutes (90%), while the other phyla, conventionally found in the digestive ecosystems of mammals, are in the minority (10%). Lachnospiraceae and Ruminococcaceae are the dominant families of the cecal ecosystem (40% and 30%) followed by Bacteroidaceae and Rikenellaceae (less than 3%) [4]. The gut microbiota, as well as the factors affecting its composition, are considered important aspects to maintaining digestive health and therefore to enhancing rabbit production [2,5]. Diet is one of the main factors influencing the microbiota and the co-occurrence patterns of the cecal bacterial community [2,5]. Specifically, dietary fat intake can modulate gut microbiota [6,7]. It has been reported that some fatty acids (FA) and in particular medium chain fatty acids (MCFA) act as bacteriostatic (growth inhibiting) or bactericidal (killing) molecules [8,9]. Then, in addition to modulating the bacterial community, the amount and type of dietary fat can have an effect on the immune function both at systemic and intestinal levels [5]. Data regarding the effect of fat type and level in rabbit diet on gut health are still very limited and lead to contradictory results [5,10].

The well-known antimicrobial resistance caused by the misuse of antibiotics drugs in animal production, and the EU ban on their in-feed use (Regulation EC/1831/2003) has led to an increase in the incidence of livestock disease and economic damage. For several years, research has dedicated great efforts to investigating alternatives able to sustain production without causing an increase of antimicrobial resistance and several natural products have been investigated [11,12].

Recently, insects have been receiving considerable attention as novel alternative feed ingredients because of their excellent nutritive properties [13] and their potential effect on animal health [14,15]. It has been shown that insects can modulate the intestinal microbiota with positive effects on poultry growth, health, and resistance against pathogens [15–19]. Lipids are a main component of insects (30%–40% of the dry matter, DM) [20] and, once extracted during the insect larvae processing, they can be used as an interesting feed ingredient in animal farming [14,15,21–26]

Black soldier fly (*Hermetia illucens*, HI) and yellow mealworm (*Tenebrio molitor*, TM) fats are characterized by a different fatty acid (FA) profile. The HI fat is rich in saturated FAs (SFAs) and in MCFAs, mainly lauric acid (C12:0; 48% of the total FA profile) [14], which has an antimicrobial effect on a wide range of microbes [27]. On the other hand, the TM fat is considered as a source of n-6 poly-unsaturated FAs (PUFAs), with high linoleic acid content (9% of the total FA profile) [14]. MCFAs are effectively absorbed and metabolized on the gastro-intestinal tract and known for their physiological and antibacterial effects on Gram positive bacteria [28–30]. Furthermore, MCFAs can exert beneficial effects on intestinal health and microbial growth inhibition [31], as well as a favorable impact on the digestive health of the growing rabbit [32]. Both the short chain FAs (SCFAs; propionic acid and butyric acid) and the MCFAs (caproic acid and caprylic acid) have a direct in vitro antimicrobial activity against *Salmonella typhimurium* [33]. Matsue et al. [27] demonstrated that lauric acid has a high antimicrobial activity against pathogenic *Bacteroides* and *Clostridium* species, and consequently can modulate intestinal health. Significant in vitro gut antimicrobial effects against D-*Streptococci* and *Lactobacilli* have also been attributed to HI prepupae fat [30]. In a recent study, Sypniewski et al. [15] showed that the addition of HI fat to replace soybean oil (S) in turkey diets significantly reduced the presence of potentially-pathogenic microbes and decreased gastrointestinal-tract (GIT) inflammation by modulating the level of IL-6 and TNF-α.

In the light of what was reported above, the aim of the present study was to evaluate the in vitro antimicrobial activities of HI and TM fats and their effects on cecal traits and gut microbiota of growing rabbits.

2. Materials and Methods

2.1. Animal Ethics Statement

The study was performed in accordance with the guidelines of the European and the Italian laws (European Directive 86 609/EEC-Italian law D.L. 116/92), and was approved by the Bioethical Committee of the University of Torino (Italy) (Ref. 386638, 4/12/2017).

2.2. In Vitro Analyses for Antimicrobial Activity of Insect Fats

Bacterial strains were selected considering their impact on rabbits and on public health [34–37]. *Salmonella tiphymurium* (CIP 60.62T), *Salmonella enteritidis* (CRBIP 19.329), *Yersinia enterocolitica* (CIP 101.776), *Pasteurella multocida* (CIP 100.610) and *Listeria monocytogenes* (CIP 82.110T) were all purchased from Institute Pasteur (Paris, France). After overnight incubation following producers' instructions, strains were mixed with TM and HI fats [14] in order to reach a final concentration of 3 Log CFU/mL, verified by immediate plate streaks. Briefly, the broths for growth were prepared as follows: 200 µL of brain heart infusion broth (BHI, Oxoid, Fisher Scientific, Rodano, Milano, Italy) were added to 250 µL of insect fat and then to 50 µL of fresh bacterial overnight culture previously quantified by measuring the optical density at 600 nm (Prixma UV/VIS 5200, Fulltech Instruments, Rome, Italy) and diluted to 4 Log CFU/ml. Only for *Pasteurella multocida*, instead of BHI, tryptic soy broth (TSB, Oxoid, Fisher Scientific, Rodano, Milano, Italy) was used. These mixes were incubated 24 hours at 37 °C with horizontal shaking (RPR horizontal rotator, LE8S, Fisher Scientific, Rodano, Milano, Italy). At regular timings (every 2 hours), three aliquots for each mix were streaked on tryptic soy agar (TSA, Oxoid, Fisher Scientific, Rodano, Milano, Italy) (only *Pasteurella multocida*) or on brain heart infusion agar (BHA, Oxoid, Fisher Scientific, Rodano, Milano, Italy). Moreover, control tests were prepared—450 µL of BHI were added with 50 µL of quantified overnight cultures to reach a final concentration of 3 Log CFU/mL of broth (A) and a mix of soybean oil (S) and bacterial broth, prepared as described above for insect fats (B). Each control and strain/fat combination were analyzed in triplicates for a total of 84 samples per bacterial species—12 samples per 6-time intervals.

2.3. Inclusion of Hermetia illucens and Tenebrio molitor Fats in the Diet of Growing Rabbits

2.3.1. Experimental Design

For this experiment, three dietary treatments were tested in 120 growing rabbits (40 rabbits/diet) from 36 to 78 days of age—a control diet (C) containing 1.5% soybean oil (S), and two diets (HID and TMD) with total replacement of S with HI and TM larvae fat, respectively. TM and HI fats were obtained from commercial companies (Ynsect, Evry, France and Hermetia Deutschland GmbH & Co. KG, Baruth/Mark, Germany) by a mechanical process using high pressure and without solvents. The three diets contained an average of 16.6% DM of crude protein (CP) and 18.6 MJ/kg DM of gross energy. Detailed information about the rabbit farming conditions and live performance are provided by Gasco et al. [14]. Ingredients and chemical composition of the diets are shown in Table 1. Briefly, no overall significant differences among experimental groups were observed for growth performance during the trial.

Table 1. Ingredients (% as fed) and chemical composition (% DM) of experimental diets. (modified from Gasco et al. [14]).

	Experimental Diets		
Ingredients	**C**	**HID**	**TMD**
Dehydrated alfalfa meal (17 g CP/100 g)	32	32	32
Alfalfa hay	7.5	7.5	7.5
Wheat bran	23.5	23.5	23.5
Barley meal	10	10	10
Dried sugar beet pulp	16	16	16
Soybean meal (44 g CP/100 g)	7	7	7
Soybean oil	1.5	-	-
Hermetia illucens fat	-	1.5	-
Tenebrio molitor fat	-	-	1.5
Cane molasses	1.2	1.2	1.2
Dicalcium phosphate	0.3	0.3	0.3
Salt	0.4	0.4	0.4
L-methionine (98 g methionine/100 g)	0.1	0.1	0.1
Vitamin-mineral premix[a]	0.5	0.5	0.5
Chemical Composition			
Dry matter, %	89.4	89.2	89.6
Ash, % DM	8.58	7.77	7.75
Crude protein, % DM	17.0	16.8	16.3
Ether extract, % DM	4.22	3.92	3.87
Neutral detergent fiber (aNDF), % DM	40.2	41.7	40.5
Acid detergent fiber (ADF), % DM	21.7	23.0	22.8
Acid detergent lignin (ADL), % DM	4.81	5.09	5.02
Gross energy, MJ/kg DM	18.50	18.50	18.62

C: control diet with soybean oil; HID: diet with *Hermetia illucens* fat; TMD: diet with *Tenebrio molitor* fat; CP: crude protein; DM: dry matter. [a]Supplementation per kilogram of feed: vitamin A, 16,000 IU; vitamin D3, 1600 IU; vitamin E acetate, 30 mg; vitamin B1, 0.8 mg; vitamin B6, 1.65 mg; niacin, 40 mg; folic acid, 1 mg; Mn, 30 mg; Fe, 116 mg; Cu, 12.5 mg; Zn, 60 mg; Co, 0.45 mg; Ca, 1.3 mg; Se, 0.3 mg.

2.3.2. Fatty Acid Profiles of Insect Lipids and Experimental Diets

The FA profiles of insect lipids and feeds were determined as reported by Gasco et al. [14]. The fatty acid methyl esters (FAME) were separated, identified, and quantified. The results were expressed as g/kg of FAME (Table 2).

Table 2. Fatty acid profiles of the dietary fats and experimental diets (g kg-1 of total FAME) (modified from Gasco et al. [14]).

Fatty acids	Dietary Fats			Experimental Diets		
	S	HI	TM	C	HID	TMD
C12:0	0.2	480	2.3	0.5	203	3.0
C14:0	0.5	103	22.2	0.9	44.7	13.3
C16:0	104	127	176	121	161	184
C18:0	44.3	19.0	23.1	28.4	20.8	22.2
BCFAs	0.1	2.9	0.8	2.4	4.1	3.4
C16:1 n-7	0.9	32.0	16.6	0.12	19.9	10.4
C18:1 n-9	230	91.1	378	201	127	273

Table 2. *Cont.*

Fatty acids	Dietary Fats			Experimental Diets		
	S	HI	TM	C	HID	TMD
C18:2 n-6	515	90.0	332	521	310	389
C18:3 n-3	70.3	10.1	18.0	74.3	62.8	55.1
SFA [1]	158	748	231	165	454	240
UFA [1]	842	252	769	835	546	760
MUFA [1]	254	141	411	236	169	309
PUFA [1]	588	111	358	599	377	451
$\sum$n3	70.5	11.7	18.3	74.7	62.8	55.1
$\sum$n6	516	91.1	333	523	311	391

S: soybean oil; HI: *Hermetia illucens* fat; TM: *Tenebrio molitor* fat; C: control diet with soybean oil; HID: diet with *Hermetia illucens* fat; TMD: diet with *Tenebrio molitor* fat; FAME: fatty acid methyl ester; BCFAs: branched-chain fatty acids; SFA: saturated fatty acid; UFA: unsaturated fatty acid; MUFA: monounsaturated fatty acid; PUFA: polyunsaturated fatty acid. [1] included minor FAs.

2.3.3. Cecal Fermentation Traits

At slaughtering (78 days of age), a total of 30 rabbits (10 animals per treatment) were randomly selected and eviscerated. The pH, cecal ammonia, and volatile FA (VFA) profiles of the cecal contents were determined as reported by Tazzoli et al. [38].

2.3.4. DNA Extraction and 16S rRNA High-Throughput Amplicon Target Sequencing

In order to observe the development and dynamics of bacterial communities, hard feces were collected from 10 rabbits per group at 36, 51, and 77 days of age, whereas the cecal contents (n = 10 per group) were taken during the slaughtering at 78 days of age. Samples from the same group and the same collection site and day were pooled together in sterilized polyethylene bags, and immediately stored at −80 °C until examination. DNA from feces and cecal samples were extracted by using a commercial kit (QIAamp Fast DNA stool Mini Kit, QIAGEN®, Hilden, Germany) following the instructions reported by the manufacturer with slightly modification. The DNA were quantified by NanoDropTM 2000 Spectrophotometer (Fisher Scientific, Rodano, Milano, Italy) and standardized at 5 ng/uL and used a template in the PCR amplifying the V3–V4 region of the 16S rRNA gene using the primers and protocols described by Klindworth et al. [39]. The PCR amplicons were cleaned according to Illumina (San Diego, CA, USA) guidelines. The sequencing was performed with a MiSeq Illumina instrument with V3 chemistry and generated 250 bp paired-end reads according to the manufacturer's instructions.

2.3.5. Bioinformatics and Statistical Analysis

After paired-end sequencing, raw reads were analyzed as previously reported by Biasato et al. [17]. Sequences were first joined using FLASH software [40] with default parameters and filtered for low quality bases (at Phred < Q20) by QIIME 1.9.0 software [41]. Reads shorter than 300 bp were discarded by using Prinseq software [42]. The USEARCH software version 8.1 was used for chimera filtering and operational taxonomic units (OTUs) were clustered at 97% of similarity threshold by UCLUST algorithms [43]. Representative sequences of each cluster were mapped against the Greengenes 16S rRNA gene database version 2013 for taxonomic assignment. In order to avoid bias due to the different sequencing depth, OTU tables were rarefied at 3996 sequences/sample. tables display the higher taxonomy resolution that was reached. When the taxonomy assignment was not able to reach the genus level, the family or phyla were displayed. Alpha diversity indices were calculated using the diversity function of the vegan package [44]. Weighted and unweight UniFrac distance matrix and OTU tables were used to find differences between samples by anosim and adonis statistical tests through the function vegan in R environment. The pairwise Wilcoxon test was used as appropriate to determine significant differences in alpha diversity or OTU abundance. A generalized linear model was used in order to test the importance of continuous or discrete variables available (sampling time and insect inclusion) on the relative abundance of bacterial genera or family. The interactions between the levels of the fixed factors were evaluated by pairwise comparisons. Not-normally distributed variables were presented as median (range interquartile) and box plots represented the interquartile range between the first and the third quartiles, with the error bars showing the lowest and the highest value. Pairwise Spearman's non-parametric correlations were used to study the relationships between the relative abundance of microbial taxa in cecal samples and VFA variables. The correlation plots were visualized in R using the coplot package of R.

The statistical analysis for data related to in vitro antimicrobial activity and cecal traits was performed using IBM SPSS Statistics V25.0.0 software (IBM, Armonk, NY, USA) by means of ANOVA, followed, if significant, by Duncan test post-hoc. Regarding the antibacterial activities of insect fats, bacterial counts was evaluated at each time point and the average results compared across the different conditions—pure bacterial broth, bacterial broth mixed with soybean oil, bacterial broth mixed with TM fat, and bacterial broth mixed with HI fat. Significance was declared at $p \leq 0.05$. A statistical trend was considered for $p \leq 0.10$. Sequences data were deposited on NCBI database under the bioproject number PRJNA645756.

3. Results

3.1. In Vitro Antimicrobial Activities of Insect Fats

Overall, the bacterial counts, when quantified broths were challenged with insect fats, were lower than the controls for three out of five pathogenic species tested, whereas only the two strains of *Salmonella* did not show any significant difference in counts when compared to control tubes counts (therefore these data are not reported in Table 3). Considering the other species, for all of them results highlighted that HI fat caused a delay in bacterial growth, implying a bacteriostatic effect (Table 3). After 24 hours of incubation, counts of *Pasteurella moultocida* broths challenged with HI fat showed a mean log difference of −4.48 and −4.76 when compared to control broths with soybean oil (control B) and no oil addition (control A), respectively. Similar results were observed for *Yersinia enterocolitica* showing a mean log difference values of −5.9 and −5.97, with control B and A, respectively. Finally, *Listeria monocytogenes* counts also led to a similar pattern of results, showing mean log difference values of −5.11 and −5.15 when comparing counts of broths challenged with HI fat and the controls B and A. All these differences were statistically significant (Table 3). On the other hand, results related to TM fat showed that only *Pasteurella multocida* was effectively inhibited in growth—the mean log difference between controls and TM-challenged broths showed values of −2.64 and −2.92, respectively, for control B and control A. These values were lower than what reported above for broth challenged with HI fat (Table 3). Finally, differences in bacterial counts between HI fat challenged broths and controls were statistically meaningful from the 4th hour of incubation only in the case of *Yersinia enterocolitica*, whereas for the other species the bacteriostatic effects were achieved only from the 8th hour of incubation. TM-fat-challenged broths showed significant effect only in *Pasteurella mutocida* after 8 hours of incubation and this was maintained until the end of the experiment (see details in Table 3).

Table 3. Distribution of average cell counts (Log (CFU)/per mL) over time for the three bacterial species for which inhibition was recorded (means ± SEM; n = 3).

Growth Conditions	T4	T6	T8	T10	T12	T24
			Pasteurella multocida			
Control (A)	4.09 (0.19)	4.14 (0.06)	4.76 (0.32)—A	4.80 (0.34)—A	5.69 (0.49)—A	6.82 (0.44)—A
Control (B)	4.11 (0.17)	3.96 (0.12)	4.17 (0.26)—A	4.60 (0.40)—A	5.44 (0.57)—A	6.54 (0.39)—A
TSB + TM fat	3.95 (0.44)	3.75 (0.39)	3.08 (0.35)—B	3.21 (0.30)—B	2.91 (0.86)—B	3.90 (1.21)—B
TSB + HI fat	3.92 (0.20)	3.48 (0.17)	3.61 (0.19)—B	2.80 (0.59)—B	2.67 (0.49)—B	2.06 (0.36)—C
ANOVA	N.S	N.S	F = 4.99; p = 0.01	F = 5.07; p = 0.01	F = 6.36; p < 0.01	F = 13.75; p < 0.01
			Yersinia enterocolitica			
Control (A)	4.36 (0.16)—A	5.19 (0.28)—A	6. 76 (0.61)—A	7.32 (0.37)—A	8.18 (0.37)—A	9.95 (0.30)—A
Control (B)	4.10 (0.21)—A	5.05 (0.37)—A	5.91 (0.41)—A	6.83 (0.48)—A	7.35 (0.68)—A	9.88 (0.33)—A
BHI + TM fat	4.05 (0.31)—A	4.39 (0.52)—AB	4.98 (0.74)—AB	5.04 (0.94)—AB	5.93 (1.41)—AB	8.09 (2.13)—A
BHI + HI fat	3.34 (0.10)—B	2.69 (0.64)—B	3.01 (1.20)—B	3.02 (1.32)—B	2.94 (1.24)—B	3.98 (2.18)—B
ANOVA	F = 3.28; p = 0.05*	F = 6.38; p < 0.01	F = 4.95; p = 0.01	F = 7.71; p < 0.01	F = 6.46; p < 0.01	F = 5.42; p < 0.01
			Listeria monocytogenes			
Control (A)	4.58 (0.35)	5.58 (0.39)	6.47 (0.41)—A	7.41 (0.41)—A	8.28 (0.46)—A	10.00 (0.26)—A
Control (B)	4.52 (0.32)	5.45 (0.42)	6.26 (0.54)—A	6.94 (0.61)—A	7.95 (0.59)—A	9.96 (0.28)—A
BHI + TM fat	4 20 (0.25)	4.59 (0.79)	5.98 (0.46)—A	6.76 (0.70)—A	7.63 (0.61)—A	9.48 (0.03)—A
BHI + HI fat	4 00 (0.92)	3.66 (0.84)	**4.1 (0.63)—B**	4.48 (0.61)—B	4.51 (0.62)—B	4.85 (0.97)—B
ANOVA	N.S	N.S	F = 3.30; p = 0.05**	F = 3.98; p = 0.03	F = 6.85; p < 0.01	F = 29.92; p < 0.01

HI: *Hermetia illucens*; TM: *Tenebrio molitor*; T: incubation time in hours; TSB: tryptic soy broth; BHI: brain heart infusion; N.S.: non-significant; SEM: standard error of the means. *Significant results (without approximation p = 0.046); **results in trend of significance (p = 0.051). In the same columns, different letters identify significantly-different results at post-hoc test (Duncan). Control (A) indicates the broth culture without oil/fat in it and control (B) indicates a broth culture grown in BHI/tryptic soy agar and soybean oil.

3.2. Cecal Fermentation Traits

The cecal fermentation traits are reported in Table 4. The supplementation of insect fat increased the total VFA contents when compared to the control group (85.3 vs. 83.9 vs. 72.4 mmol/L; $p < 0.05$), whereas the pH, the cecal ammonia-N content, and the molar proportion of the different VFAs and their ratio were not influenced by the supplementation of insect fats ($p > 0.05$).

Table 4. Effect of lipid sources on cecal traits and fermentation parameters (n = 10 rabbits/group).

	C	HID	TMD	SEM	*p*-Value
pH	6.1	5.9	5.9	0.01	0.15
N–NH3 (mmol/L)	2.2	3.0	3.1	0.23	0.25
Total VFA (mmol/L)	72.4b	85.3a	83.9a	2.31	0.03
Acetic acid (C2; mmol/100 mmol VFA)	77.8	78.1	76.6	0.43	0.30
Propionic acid (C3; mmol/100 mmol VFA)	5.3	5.0	5.4	0.19	0.71
Butyric acid (C4; mmol/100 mmol VFA)	16.2	16.1	17.2	0.42	0.53
Valeric acid (C5; mmol/100 mmol VFA)	0.5	0.4	0.5	0.03	0.59
Caproic acid (C6; mmol/100 mmol VFA)	0.3	0.3	0.3	0.03	0.45
C2/C3 ratio	15.2	16.0	14.6	0.54	0.58
C2/C4 ratio	4.9	5.0	4.6	0.16	0.63

C: control diet with soybean oil; HID: diet with *Hermetia illucens* fat; TMD: diet with *Tenebrio molitor* fat; SEM: standard error of the means; N–NH3: ammonia-N; VFA: volatile fatty acid.

3.3. Cecal and Fecal Microbiota Characterization

The total number of high-quality paired-end sequences obtained from 16S rRNA sequencing reached 13,448,661 raw reads. After the filtering, 7,801,336 reads passed the filters applied through QIIME, with a median value of 59,114 ± 52,946 reads/sample, and a mean sequence length of 441 bp. The rarefaction analysis and Good's coverage, expressed as a median percentage (93%), also indicated satisfactory coverage for all the samples. We applied a generalized linear model (GLM) for the alpha-diversity in the fecal samples in order to test the effect of the treatment across time. The Chao1 index increased ($p < 0.01$) in the TM and HI groups when compared to the C group ($p < 0.05$), while the Shannon index was affected by the sampling time only. However, the number of observed species significantly increased in the HI group in comparison with the other diets ($p < 0.05$) (Figure 1).

No significant differences were observed when comparing the alpha diversity index as a function of the different diets in the cecal samples. Going more deeply in the microbiota comparison, adonis and analysis of similarity (ANOSIM) statistical tests, based on the weighted and unweighted UniFrac distance matrix showed significant differences among samples as a function of the sampling time ($p < 0.05$). However, the dietary inclusion of insect fat did not show any significant effect in the microbiota composition of the cecal samples.

Figure 2 summarizes the distribution of the microbiota across time and samples and displays a simple microbiota composition dominated by the presence of *Bacteroides*, Clostridiales, Lachnospiraceae, Ruminococcaceae and *Ruminococcus*. Comparing the relative abundance of the main OTUs across the fecal samples grouped as a function of the dietary supplementation by the GLM function, it was possible to observe that HI inclusion level increased the relative abundance of *Akkermansia* ($p < 0.05$) and *Ruminococcus* ($p < 0.01$) and reduced the presence of *Citrobacter* ($p < 0.05$).

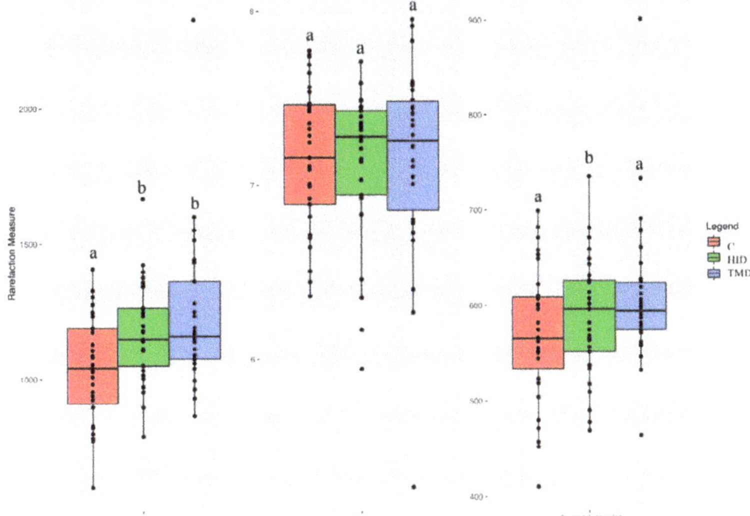

Figure 1. Boxplots showing the alpha diversity rarefaction index across fecal samples of rabbits fed with TMD (blue box), HID (green box), or C (red box) diets; C: control diet with soybean oil; HID: diet with *Hermetia illucens* fat; TMD: diet with *Tenebrio molitor* fat; Boxes represent the interquartile range (IQR) between the first and third quartile and the line inside represents the median (2nd quartile). Whiskers denote the lowest and the highest values within 1.56 IQR from the first and third quartile, respectively. Different letters in each box indicate significant difference.

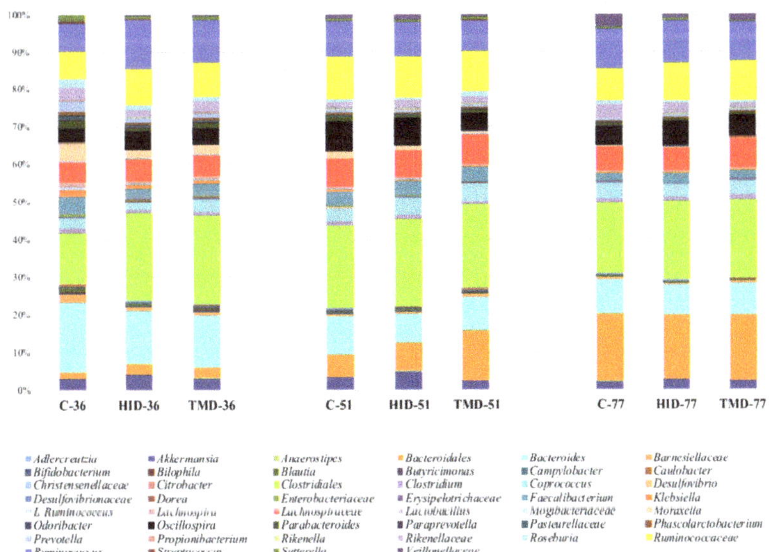

Figure 2. Taxonomic groups detected in fecal samples by means of 16S amplicon target sequencing. Only operational taxonomic units (OTUs) with an incidence above 0.2% in at least two samples are shown. The samples are labelled according to the type of fat supplementation (C: control diet with soybean oil; HID: diet with *Hermetia illucens* fat; TMD: diet with *Tenebrio molitor* fat) and sampling time (36, 51, or 77 days).

Both HID and TMD increased the relative abundance of Clostridiales ($p < 0.01$) and Desulfovibrionaceae ($p < 0.01$), while reducing the relative abundance of *Lachnospira* ($p < 0.05$) and *Phascolarctobacterium* ($p < 0.01$) when compared to the C diet (Figure 3).

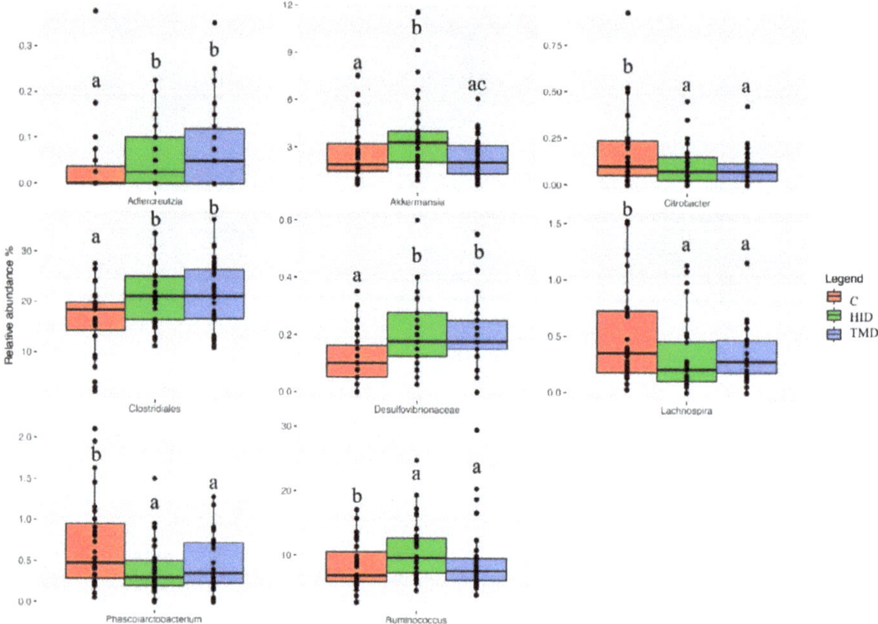

Figure 3. Boxplots showing the relative abundance at genus or family level of operational taxonomic units (OTUs) differential abundant based on the generalized linear model (GLM) test in fecal samples of rabbits fed with TMD (blue box), HID (green box), or C diets (red box). Different letters in each box indicate significant difference; C: control diet with soybean oil; HID: diet with *Hermetia illucens* fat; TMD: diet with *Tenebrio molitor* fat.

Taking into account the cecal samples (Figure 4), no differences were observed regarding the microbial composition and distribution. However, the pairwise comparisons using the Wilcoxon rank sum test showed that TM inclusion reduced the relative abundance of *Klebsiella*, *Lachnospira*, *Parabacteroides*, and *Odoribacter* compared to the C and HID (Figure 5, $p < 0.05$).

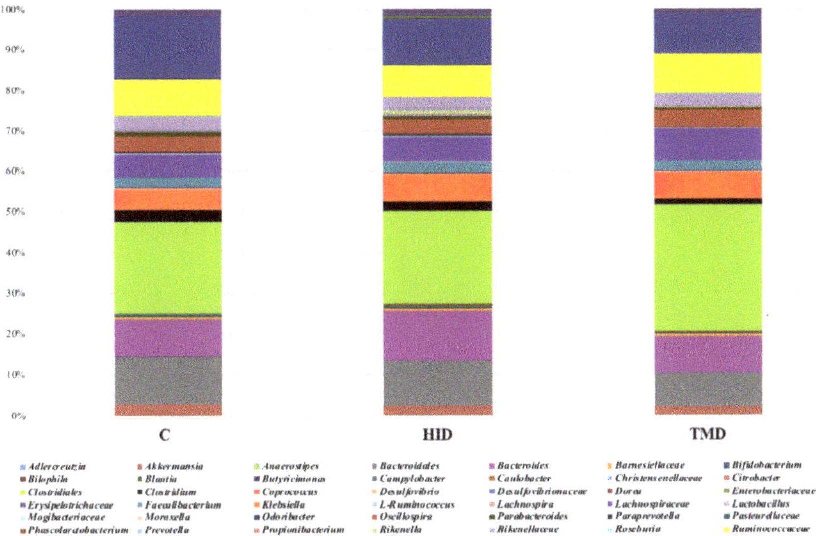

Figure 4. Taxonomic groups detected in cecal samples by means of 16S amplicon target sequencing. Only operational taxonomic units (OTUs) with an incidence above 0.2% in at least two samples are shown. The samples are labelled according to the type of fat supplementation, i.e., TMD, HID, or C diets; C: control diet with soybean oil; HID: diet with *Hermetia illucens* fat; TMD: diet with *Tenebrio molitor* fat.

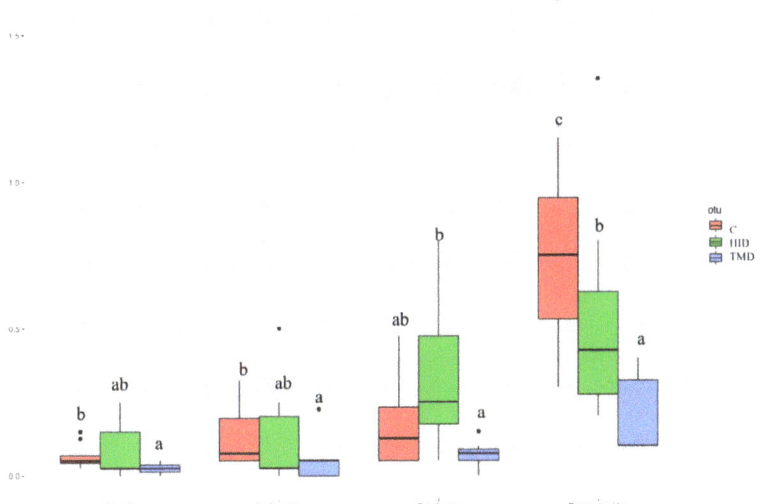

Figure 5. Boxplots showing the relative abundance at genus or family level of the operational taxonomic units (OTUs) differential abundance based on Wilcoxon rank sum test in cecal samples of rabbits fed with TM (blue box), HI (green box), or C (red box) diets. Different letters in each box indicate significant difference; C: control diet with soybean oil; HID: diet with *Hermetia illucens* fat; TMD: diet with *Tenebrio molitor* fat.

The correlations between VFA contents and microbiota are summarized in Figure 6. In detail, strong positive correlations between *Lachnospira* and propanoic acid, *Akkermansia*, Clostridiales, and NH3-N, and *Phascolarctobacterium* and acetic, propanoic, and caproic acids were detected ($p < 0.05$).

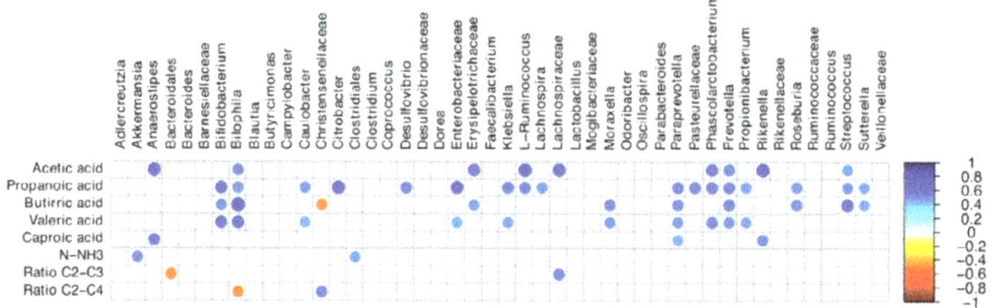

Figure 6. Spearman's rank correlation matrix of cecal operational taxonomic unit (OTU) abundance and the volatile fatty acid profile. The colors of the scale bar denote the nature of the correlation, with 1 indicating a positive correlation (blue) and -1 indicating a negative correlation (dark red), between the two datasets; N–NH3: ammonia-N; C2: Acetic acid; C3: Propanoic acid; C4: Butyric acid.

4. Discussion

The current study was conducted to evaluate the in vitro antimicrobial properties of HI and TM fats and their effect as an alternative lipid sources in rabbit diets. The possible utilization of insect fats in animal diets has been investigated recently, and so far, a few papers focusing on antimicrobial activities in vitro or in vivo are available, but no data are present in the literature on pathogen growth rate in presence of insect fats. Mustafa et al. [45] showed that the oil extracted from the melon bug (*Aspongopus vidulatus*) was able to inhibit the growth of bacterial species by using the agar well diffusion method, with only Gram positive bacteria (*Staphylococcus, Bacillus,* and *Enterococcus*) being susceptible to crude oil extracts. *Salmonella paratyphi* was also tested and no inhibition was recorded, similarly to what was observed in the present research with other serovars. A recent study of Spranghers et al. [30] on the in vitro effects of fats extracted from HI (from prepupae fed to weaned piglets) pointed out that D-*Streptococci* and *Lactobacilli* were the only bacterial populations that reduced their load after being challenged with insect fats, whereas no effects were recorded on Gram negative bacteria. The data presented in our paper suggest that fats extracted from TM and HI are able to delay bacterial growth of both Gram positive and Gram negative pathogenic bacteria, even if the susceptibility changes with the considered species. Interestingly, no bactericidal effect was observed, thus indicating the possibility of bacterial cells repairing damage that may be induced from FAs or monoglycerides [8]. The antimicrobial effect of TM fat was lower than that of HI fat. These results may be related to the different concentrations of SFA. Indeed, the HI fat used in this study was composed by 79% of SFA (25% in TM fat), and the major component was lauric acid, already reported as very effective against many bacterial species [8,30]. In a systematic study of the in vitro antimicrobial activities of FAs, monoglycerides, and diglycerides, Kabara et al. [46] showed that lauric acid was the most active FA against gram-positive bacteria. On the other hand, the antibacterial effects of unsaturated fatty acids (UFA) reported in literature [8] were not detected in this study, probably in relation to the prolonged incubation time and the temperature of 37 °C, which may have been responsible for the reduction/impairment of the activity of these molecules that characterize TM fat (75% of the FA vs. 21% of HI fat). This hypothesis may also explain why soybean oil did not show any activity, considering its higher level in monounsaturated fatty acids (MUFAs) and PUFAs among all fats used in this work.

The present observations highlighted the possibility of using insect fats in feed formulation, considering that they may be important for controlling growth of important microbial pathogens such as *Listeria monocytogenes, Yersinia enterocolitica,* and *Pasteurella multocida* that may be important pathogens for rabbits [34,37] or part of the gut flora potentially contaminating rabbit meat during slaughtering [35,36]. Reducing microbial loads in the rabbit gut, apart from animal welfare and safety implications, may also be important for food safety management during slaughtering operations. Studies performed on piglets fed with diets rich in FAs showed a decreased load of microbes in the stomach and gut, mostly if undesired species were considered (i.e., *streptococci*), thus promoting gut heath and growth performances [30,47]. However, as already emphasized by Spranghers et al. [30], more studies need to be performed in vivo, in order to assess the activity of fat in the gut, considering that the bacterial activities and the digestive enzymatic systems of the rabbit (i.e., lipases) may neutralize the FAs, therefore limiting their activity. Furthermore, the presence of FAs as antimicrobial molecules in feed may prevent the use of classical antibiotics in meat production, limiting the diffusion of microbes harboring antimicrobial resistance genes. In fact, antimicrobial resistance mechanisms were rarely observed in microbial species challenged with FAs [47], and to date no genetic modifications induced by exposing bacterial cells to antimicrobial FAs have been shown [8,48].

As far as the in vivo trial is concerned, there was a lack of differences among groups in cecal fermentation traits in our study. Cecal pH and VFA content are the main variables characterizing the extent and the pattern of cecal fermentation, thus constituting an indirect estimate and an important tool to qualitatively evaluate the cecal microbial activity. The dietary HI or TM fat inclusion led to a greater total VFA content in the cecum than that of control diet, thus potentially enhancing the gut with a modification of the fermentation patterns and the composition of the cecal microflora. Peeters et al. [49] previously observed that a high concentration of total cecal VFA in rabbits had a protective effect against enteropathogenic *Escherichia coli* infection. However, the molar proportion of different VFA was not affected by the total replacement of soybean oil with HI and TM larvae fats.

The present study is the first to investigate gut microbiota of growing rabbits fed diets supplemented with insect fats. As reported previously, HI fat was rich in MCFAs which have a higher antimicrobial activity and thus can stimulate gastrointestinal tract health through inhibition of potentially-pathogenic bacteria [31,33]. MCFAs, and in particular lauric acid, are presumably absorbed in the upper digestive tract, and might be useful in protecting against microbial infection, modulating inflammation, healing wounds, and controlling the balance and distribution of bacteria in gut microbiota [50]. The present study is the first to investigate gut microbiota of growing rabbits fed diets supplemented with insect fats. The results of this study revealed an enrichment of different taxa according to the dietary treatment and a similar microbial diversity and richness between feces and cecum samples. Cecal microbiota is a primary determinant for rabbit health, whereas the fecal microbiota provides an accurate method for studying the evolution of rabbit gut microbiota from weaning to slaughtering [51,52]. Investigating the differences between rabbits fed the C and the HI and TM fat diets, in the current study, no differences were found with regard to alpha-diversity measures in cecal and fecal samples. However, we observed a higher microbial diversity in the HI and TM groups. High levels of diversity generally help intestinal microbiota to determine effective colonization resistance against potential invading pathogens and to modulate animal reaction after a stress-environment [53]. Based on such considerations, the above-mentioned findings are indicative of a positive HI and TM fat-related effect on the gut microbiota of rabbits. *Firmicutes, Bacteroidetes,* and *Verrucomicrobia* represented the dominant bacterial phyla in the control and insect-fat-fed rabbits of the present study. These findings overall agree with previous research that identified *Firmicutes* and *Bacteroides* as the main bacterial phyla in the gut microbiota of rabbits [3,52,54,55]. In relation to the genera composition, the *Bacteroides*, Bacteroidales, Clostridiales, Lachnospiraceae, Ruminococcaceae, and *Ruminococcus*, mainly colonized the cecal and fecal microbiota of the rabbits fed soybean oil or insect fats in the current study. These findings are also in agreement with previous studies, which observed *Bacteroides* [52,55], Clostridiales [54], and *Ruminococcus* [53,56] as being the main bacterial genera in cecal and fecal microbiota of rabbits and being capable of degrading the polysaccharide and amino acid fermentation to produce VFAs. These findings could be the reason for its overrepresentation in cecum where it is supposed to play an active role [52,57]. A previous study reported that the presence of this taxa was positively correlated with the feed conversion rate [58].

Regarding the microbial composition, we did not observe any strong effects as a consequence of the dietary inclusion of HI and TM fats. However, a signature in the microbial population was observed. The fat of TM reduced some taxa such as *Klebsiella, Lachnospira, Parabacteroides,* and *Odoribacter*. On the other hand, the dietary supplementation of HI and TM fats enriched the presence of Clostridiales, Desulfovibrionaceae, *Ruminococcus,* and *Akkermansia*, which are the main taxa in the gut microbiota of rabbits [51]. It is well reported that *Akkermansia* can be considered a new-generation probiotic, able to degrade the mucin in the gut with the production of beneficial molecules like SCFAs thus exerting a significant improvement in the gut barrier and in the maintenance of intestinal health [59–61]. In addition, it was suggested that *Akkermansia* have an important role in the hydrolysis of various dietary polysaccharides, contributing to increase cellulose digestibility as well as methane metabolism [62–64]. *Akkermansia* would be also involved in carbohydrate digestion and in immune protection against inflammation [54]. The increase of this taxon related with the insect fat inclusion suggests an optimal gut environment in our rabbits, even if this observation needs further investigation to be confirmed. In addition, a strong positive correlation between *Akkermansia* and NH3-N was observed. This result confirmed that gut microbiota is responsible for a variety of metabolic activities including production of biologically-active substances. The Ruminococcaceae family is considered as an important producer of short-chain FAs (mainly butyrate, acetic, and succinic acids) through glucose metabolism and cellulose digestion [65,66]. It was reported that members of Ruminococcaceae are important components of the beneficial microbiota of several herbivores [67,68]. Their presence is related to an improvement of the immune system of the host via intestinal mucus degradation and a prevention of acidosis via lactate degradation [69]. Ruminococcaceae has been associated with antibiotic biosynthesis in a metatranscriptomic study on the human gut microbiota, suggesting a role in gut health [70]. The supplementation of HI fat also increased the presence of *Ruminococcus* (belonging to Lachnospiraceae family), that are butyrate-producing bacteria. This ability was also confirmed by the strong positive correlation between L-*Ruminococcus*, acetic, and propionic acid. The identification of short-chain FA-producing bacteria allows us to hypothesize a similar way of action for HI fat in the rabbit's gut. Therefore, the increase in the above-mentioned bacterial taxa by dietary HI fat may have helped the rabbits to maintain a healthy gut. The microbial signature driven by the inclusion of insect fat increases the presence of *Bacteroides, Clostridium, Akkermansia,* and *Ruminococcus* and suggests that dietary HI and TM fats may exert a positive influence on the cecal microbiota of rabbits. It should be pointed out that rabbits fed HI and TM fats showed no significant alterations at histopathological level and no differences in growth performance [14]. However, since a clear cause–effect relationship between diversity and composition of cecal microbiota and rabbit performance has not yet been confirmed, the gut microbiota findings need to be contextualized to underline the effect of insect fats in rabbit diets.

5. Conclusions

In conclusion, the results of the present study provide new information about the in vitro antibacterial properties and the in vivo effects at the gut level of the replacement of soybean oil with HI and TM larvae fat in rabbit diets. The in vitro activities of HI or TM fats against *Pasteurella, Yersinia*, and known pathogens of the rabbit gut, indicate a potential for impairing their growth in vivo in rabbits. Furthermore, the dietary inclusion of HI and TM fats stimulate VFA production at the cecum and may positively modulate the cecal and fecal microbiota of growing rabbits. However, further research is needed to confirm the antimicrobial potentiality of insect fats in rabbit feeding.

Author Contributions: Conceptualization, S.D., L.G., A.S., A.T., G.X., and F.G.; data curation, I.F., D.M.N., A.T., G.X., S.D., S.M., D.S., and A.C.B.; funding acquisition L.G.; methodology, S.D., D.M.N., I.F., F.G., I.B., L.C., and L.G.; writing—original draft preparation, S.D., I.F., and D.M.N.; writing—review and editing, all the authors. All authors have read and agreed to the published version of the manuscript.

Funding: Financial support for this work was provided by Cassa Risparmio Torino grant (GASL_RIC_N_COMP_18_02) and by the University of Turin (ex 60%) grants (BIAI_RILO_19_01).

Acknowledgments: For Schiavone A. this work is the result of the visit (21142/IV/19) funded by the Fundación Séneca-Agencia de Ciencia y Tecnología de la Región de Murcia (Spain) in connection with the "Jiménez De La Espada" Regional Programme for Mobility, Collaboration and Knowledge Exchange (Spain).

Conflicts of Interest: The authors declare no conflict of interest. The funders had no role in the design of the study; in the collection, analyses, or interpretation of data; in the writing of the manuscript, or in the decision to publish the results.

References

1. Licois, D.; Wyers, M.; Coudert, P. Epizootic Rabbit Enteropathy: Experimental transmission and clinical characterization. *Vet. Res.* **2005**, *36*, 601–613. [CrossRef] [PubMed]
2. Combes, S.; Fortun-Lamothe, L.; Cauquil, L.; Gidenne, T. Engineering the rabbit digestive ecosystem to improve digestive health and efficacy. *Animal* **2013**, *7*, 1429–1439. [CrossRef] [PubMed]
3. Vántus, V.B.; Kovács, M.; Zsolnai, A. The rabbit cecal microbiota: Development, composition and its role in the prevention of digestive diseases—A review on recent literature in the light of molecular genetic methods. *Acta Agrar. Kaposváriensis* **2014**, *18*, 55–65.
4. Fortun-Lamothe, L.; Theau-Clément, M.; Combes, S.; Allain, D.; Lebas, F.; Le Normand, B.; Gidenne, T. Physiologie. In *Le Lapin de la Biologie à L'élevage*, 1st ed.; Gidenne, T., Ed.; Quae: Versailles, France, 2015; pp. 48–49. ISBN 9782759224166.
5. Fortun-Lamothe, L.; Boullier, S. A review on the interactions between gut microflora and digestive mucosal immunity. Possible ways to improve the health of rabbits. *Livest. Sci.* **2007**, *107*, 1–18. [CrossRef]
6. Kim, S.J.; Kim, S.E.; Kim, A.R.; Kang, S.; Park, M.Y.; Sung, M.K. Dietary fat intake and age modulate the composition of the gut microbiota and colonic inflammation in C57BL/6J mice. *BMC Microbiol.* **2019**, *19*, 193. [CrossRef]
7. Yang, Q.; Liang, Q.; Balakrishnan, B.; Belobrajdic, D.P.; Feng, Q.J.; Zhang, W. Role of dietary nutrients in the modulation of gut microbiota: A narrative review. *Nutrients* **2020**, *12*, 381. [CrossRef]
8. Yoon, B.K.; Jackman, J.A.; Valle-González, E.R.; Cho, N.J. Antibacterial free fatty acids and monoglycerides: Biological activities, experimental testing, and therapeutic applications. *Int. J. Mol. Sci.* **2018**, *19*, 1114. [CrossRef]
9. Xiccato, G. Fat Digestion. In *Nutrition of the Rabbit*, 3rd ed.; De Blas, C., Wiseman, J., Eds.; CAB International: Wallingford, UK, 2020; pp. 58–68. ISBN 9781789241273. [CrossRef]
10. Gidenne, T.; Lebas, F.; Licois, D.; García, J. Nutrition and Feeding Strategy: Impact on Health Status. In *Nutrition of the Rabbit*, 3rd ed.; De Blas, C., Wiseman, J., Eds.; CAB International: Wallingford, UK, 2020; pp. 193–221. ISBN 9781789241273. [CrossRef]
11. Christaki, E.; Marcou, M.; Tofarides, A. Antimicrobial resistance in Bacteria: Mechanisms, evolution, and persistence. Review. *J. Mol. Evol.* **2020**, *88*, 26–40. [CrossRef]
12. Girard, M.; Bee, G. Invited review: Tannins as a potential alternative to antibiotics to prevent coliform diarrhea in weaned pigs. *Animal* **2020**, *14*, 95–107. [CrossRef]
13. Gasco, L.; Biasato, I.; Dabbou, S.; Schiavone, A.; Gai, F. Animals fed insect-based diets: State-of-the-art on digestibility, performance and product quality. *Animals* **2019**, *9*, 170. [CrossRef]
14. Gasco, L.; Dabbou, S.; Trocino, A.; Xiccato, G.; Capucchio, M.T.; Biasato, I.; Dezzutto, D.; Birolo, M.; Meneguz, M.; Schiavone, A.; et al. Effect of dietary supplementation with insect fats on growth performance, digestive efficiency and health of rabbits. *J. Anim. Sci. Biotechnol.* **2019**, *10*, 4. [CrossRef] [PubMed]
15. Sypniewski, J.; Kierończyk, B.; Benzertiha, A.; Mikołajczak, Z.; Pruszyńska Oszmałek, E.; Kołodziejski, P.; Sassek, M.; Rawski, M.; Czekała, W.; Józefiak, D. Replacement of soybean oil by *Hermetia illucens* fat in turkey nutrition: Effect on performance, digestibility, microbial community, immune and physiological status and final product quality. *Br. Poult. Sci.* **2020**, *61*, 294–302. [CrossRef]
16. Borrelli, L.; Coretti, L.; Dipineto, L.; Bovera, F.; Menna, F.; Chiariotti, L.; Nizza, A.; Lembo, F.; Fioretti, A. Insect based diet, a promising nutritional source, modulates gut microbiota composition and SCFAs production in laying hens. *Sci. Rep.* **2017**, *7*, 16269. [CrossRef]

17. Biasato, I.; Ferrocino, I.; Biasibetti, E.; Grego, E.; Dabbou, S.; Sereno, A.; Gai, F.; Gasco, L.; Schiavone, A.; Cocolin, L.; et al. Modulation of intestinal microbiota, morphology and mucin composition by dietary insect meal inclusion in free-range chickens. *BMC Vet. Res.* **2018**, *14*, 383. [CrossRef]
18. Biasato, I.; Ferrocino, I.; Grego, E.; Dabbou, S.; Gai, F.; Gasco, L.; Cocolin, L.; Capucchio, M.T.; Schiavone, A. Gut microbiota and mucin composition in female broiler chickens fed diets including yellow mealworm (*Tenebrio molitor*, L.). *Animals*, 2019; 9, 213. [CrossRef]
19. Biasato, I.; Ferrocino, I.; Dabbou, S.; Evangelista, R.; Gai, F.; Gasco, L. Black soldier fly and gut health in broiler chickens: Insights into the relationship between caecal microbiota and intestinal mucin composition. *J. Anim. Sci. Biotechnol.* **2020**, *11*, 11. [CrossRef] [PubMed]
20. Berezina, N. Insects: Novel source of lipids for a fan of applications. *Oil Seeds Fats Crop. Lipids.* **2017**, *24*, D402. [CrossRef]
21. Schiavone, A.; Cullere, M.; De Marco, M.; Meneguz, M.; Biasato, I.; Bergagna, S.; Dezzutto, D.; Gai, F.; Dabbou, S.; Gasco, L.; et al. Partial or total replacement of soybean oil by black soldier larvae (*Hermetia illucens* L.) fat in broiler diets: Effect on growth performances, feed-choice, blood traits, carcass characteristics and meat quality. *Ital. J. Anim. Sci.* **2017**, *16*, 93–100. [CrossRef]
22. Schiavone, A.; Dabbou, S.; De Marco, M.; Cullere, M.; Biasato, I.; Biasibetti, E.; Capucchio, M.T.; Bergagna, S.; Dezzutto, D.; Meneguz, M.; et al. Black soldier fly (*Hermetia illucens* L.) larva fat inclusion in finisher broiler chicken diet as an alternative fat source. *Animal* **2018**, *12*, 2032–2039. [CrossRef]
23. Dalle Zotte, A.; Cullere, M.; Martins, C.; Alves, S.P.; Freire, J.P.B.; Falcão-e-Cunha, L.; Bessa, R.J.B. Incorporation of Black Soldier Fly (*Hermetia illucens* L.) larvae fat or extruded linseed in diets of growing rabbits and their effects on meat quality traits including detailed fatty acid composition. *Meat Sci.* **2018**, *146*, 50–56. [CrossRef]
24. Kierónczyk, B.; Rawski, M.; Józefiak, A.; Mazurkiewicz, J.; Świątkiewicz, S.; Siwek, M.; Bednarczyk, M.; Szumacher-Strabel, M.; Cieślak, A.; Benzertiha, A.; et al. Effects of replacing soybean oil with selected insect fats on broilers. *Anim. Feed Sci. Technol.* **2018**, *240*, 170–183. [CrossRef]
25. Benzertiha, A.; Kierónczyk, B.; Rawski, M.; Kołodziejski, P.; Bryszak, M.; Józefiak, D. Insect oil as an alternative to palm oil and poultry fat in broiler chicken nutrition. *Animals* **2019**, *9*, 116. [CrossRef]
26. Gasco, L.; Dabbou, S.; Gai, F.; Brugiapaglia, A.; Schiavone, A.; Birolo, M.; Xiccato, G.; Trocino, A. Quality and consumer acceptance of meat from rabbits fed diets in which soybean oil is replaced with black soldier fly and yellow mealworm fats. *Animals* **2019**, *9*, 629. [CrossRef]
27. Matsue, M.; Mori, Y.; Nagase, S.; Sugiyama, Y.; Hirano, R.; Ogai, K.; Ogura, K.; Kurihara, S.; Okamoto, S. Measuring the antimicrobial activity of lauric acid against various bacteria in human gut microbiota using a new method. *Cell Transplant.* **2019**, *28*, 1528–1541. [CrossRef]
28. Skřivanová, E.; Marounek, M.; Benda, V.; Brezina, P. Susceptibility of *Escherichia coli*, *Salmonella* s.p. and *Clostridium perfringens* to organic acids and monolaurin. *Vet. Med.* **2006**, *51*, 81–88. [CrossRef]
29. Zeitz, J.O.; Fennhoff, J.; Kluge, H.; Stangl, G.I.; Eder, K. Effects of dietary fats rich in lauric and myristic acid on performance, intestinal morphology, gut microbes, and meat quality in broilers. *Poult. Sci.* **2015**, *94*, 2404–2413. [CrossRef] [PubMed]
30. Spranghers, T.; Michiels, J.; Vrancx, J.; Ovyn, A.; Eeckhout, M.; De Clercq, P.; De Smet, S. Gut antimicrobial effects and nutritional value of black soldier fly (*Hermetia illucens* L.) prepupae for weaned piglets. *Anim. Feed Sci. Tech.* **2018**, *235*, 33–42. [CrossRef]
31. Baltić, B.; Starčević, M.; Đorđević, J.; Mrdović, B.; Marković, R. Importance of medium chain fatty acids in animal nutrition. *IOP Conf. Ser. Earth Environ. Sci.* **2017**, *85*, 012048. [CrossRef]
32. Skřivanová, V.; Marounek, M. A note on the effect of triacylglycerols of caprylic and capric fatty acid on performance, mortality, and digestibility of nutrients in young rabbits. *Anim. Feed Sci. Tech.* **2006**, *127*, 161–168. [CrossRef]
33. Boyen, F.; Haesebrouck, F.; Vanparys, A.; Volf, J.; Mahu, M.; Van Immerseel, F.; Rychlik, I.; Dewulf, J.; Ducatelle, R.; Pasmans, F. Coated fatty acids alter virulence properties of *Salmonella Typhimurium* and decrease intestinal colonization of pigs. *Vet. Microbiol.* **2008**, *132*, 319–327. [CrossRef]
34. Rodriguez-Calleja, J.M.; Garcia-Lopez, I.; Garcia Lopez, M.L.; Santos, J.A.; Otero, A. Rabbit meats as a source of bacterial foodborne pathogens. *J. Food Prot.* **2006**, *69*, 1106–1112. [CrossRef]
35. De Cesare, A.; Parisi, A.; Mioni, R.; Comin, D.; Lucchi, A.; Manfreda, G. *Listeria monocytogenes* circulating in rabbit meat products and slaughterhouses in Italy: Prevalence data and comparison among typing results. *Foodborne Pathog. Dis.* **2017**, *14*, 167–176. [CrossRef] [PubMed]

36. Kylie, K.; McEwen, S.A.; Boerlin, P.; Reid-Smith, M.J.; Weese, J.S.; Turner, P.V. Prevalence of antimicrobial resistance in fecal *Escherichia coli* and *Salmonella enterica* in Canadian commercial meat, companion, laboratory, and shelter rabbits (*Oryctolagus cuniculus*) and its association with routine antimicrobial use in commercial meat rabbits. *Prev. Vet. Med.* **2017**, *147*, 53–57. [PubMed]
37. Massacci, F.R.; Magistrali, C.F.; Cucco, L.; Curcio, L.; Bano, L.; Mangili, P.M.; Scoccia, E.; Bisgaard, M.; Aalbæk, B.; Christensen, H. Characterization of *Pasteurella multocida* involved in rabbit infections. *Vet. Microbiol.* **2018**, *213*, 66–72. [CrossRef] [PubMed]
38. Tazzoli, M.; Trocino, A.; Birolo, M.; Radaelli, G.; Xiccato, G. Optimizing feed efficiency and nitrogen excretion in growing rabbits by increasing dietary energy with high-starch, high-soluble fibre, low-insoluble fibre supply at low protein levels. *Livest. Sci.* **2015**, *172*, 59–68. [CrossRef]
39. Klindworth, A.; Pruesse, E.; Schweer, T.; Peplies, J.; Quast, C.; Horn, M.; Glöckner, F.O. Evaluation of general 16S ribosomal RNA gene PCR primers for classical and next-generation sequencing-based diversity studies. *Nucleic Acids Res.* **2013**, *41*, e1. [CrossRef]
40. Magoč, T.; Salzberg, S.L. FLASH: Fast length adjustment of short reads to improve genome assemblies. *Bioinformatics* **2011**, *27*, 2957–2963. [CrossRef]
41. Caporaso, J.G.; Kuczynski, J.; Stombaugh, J.; Bittinger, K.; Bushman, F.D.; Costello, E.K.; Fierer, N.; Peña, A.G.; Goodrich, J.K.; Gordon, J.I.; et al. QIIME allows analysis of high-throughput community sequencing data. *Nat. Methods* **2010**, *7*, 335–336. [CrossRef]
42. Schmieder, R.; Edwards, R. Quality control and preprocessing of metagenomic datasets. *Bioinformatics* **2011**, *27*, 863–864. [CrossRef]
43. Edgar, R.C. Search and clustering orders of magnitude faster than BLAST. *Bioinformatics* **2010**, *26*, 2460–2461. [CrossRef]
44. Dixon, A.L.; Liang, L.; Moffatt, M.F.; Chen, W.; Heath, S.; Wong, K.C.C.; Taylor, J.; Burnett, E.; Gut, I.; Farrall, M.; et al. A genome-wide association study of global gene expression. *Nat. Genet.* **2007**, *39*, 1202–1207. [CrossRef]
45. Mustafa, N.E.M.; Mariod, A.A.; Matthäus, B. Antibacterial activity of *Aspongopus Viduatus* (melon bug) oil. *J. Food Saf.* **2008**, *28*, 577–586. [CrossRef]
46. Kabara, J.J.; Swieczkowski, D.M.; Conley, D.J.; Truant, J.P. Fatty Acids and Derivatives as Antimicrobial Agents. *Antimicrob. Agents Chemother.* **1972**, *2*, 23–28. [CrossRef] [PubMed]
47. Desbois, A.P. Potential applications of antimicrobial fatty acids in medicine, agriculture and other industries. *Recent Pat. Antiinfect. Drug Discov.* **2012**, *7*, 111–122. [CrossRef] [PubMed]
48. Davies, J.; Davies, D. Origins and evolution of antibiotic resistance. *Microbiol. Mol. Biol. Rev.* **2010**, *74*, 417–433. [CrossRef] [PubMed]
49. Peeters, J.E.; Maertens, L.; Orsenigo, R.; Colin, M. Influence of dietary beet pulp on caecal VFA, experimental colibacillosis and iota-enterotoxaemia in rabbits. *Anim. Feed Sci. Tech.* **1995**, *51*, 123–139. [CrossRef]
50. Dayrit, F.M. The properties of lauric acid and their significance in coconut oil. *J. Am. Oil Chem. Soc.* **2015**, *92*, 1–15. [CrossRef]
51. Kylie, K.; Weese, J.S.; Turner, P.V. Comparison of the fecal microbiota of domestic commercial meat, laboratory, companion, and shelter rabbits (*Oryctolagus cuniculi*). *BMC Vet. Res.* **2018**, *14*, 143. [CrossRef]
52. Fang, S.; Chen, X.; Zhou, L.; Wang, C.; Chen, Q.; Lin, R.; Xiao, T.; Gan, Q. Faecal microbiota and functional capacity associated with weaning weight in meat rabbits. *Microb. Biotechnol.* **2019**, *12*, 1441–1452. [CrossRef]
53. Sudo, N. Role of gut microbiota in brain function and stress-related pathology. *Biosci. Microbiota. Food. Health* **2019**, *38*, 75–80. [CrossRef]
54. Velasco-Galilea, M.; Piles, M.; Vinas, M.; Rafel, O.; Gonzalez-Rodriguez, O.; Guivernau, M.; Sánchez, J.P. Rabbit microbiota changes throughout the intestinal tract. *Front. Microbiol.* **2018**, *9*, 2144. [CrossRef]
55. Chen, S.Y.; Deng, F.; Jia, X.; Liu, H.; Zhang, G.W.; Lai, S.J. Gut microbiota profiling with differential tolerance against the reduced dietary fibre level in rabbit. *Sci. Rep.* **2019**, *9*, 1–9. [CrossRef] [PubMed]
56. Dabbou, S.; Ferrocino, I.; Kovitvadhi, A.; Dabbou, S.; Bergagna, S.; Dezzuto, D.; Schiavone, A.; Cocolin, L.; Gai, F.; Santoro, V.; et al. Bilberry pomace in rabbit nutrition: Effects on growth performance, apparent digestibility, caecal traits, bacterial community and antioxidant status. *Animal* **2019**, *13*, 53–63. [CrossRef] [PubMed]
57. Dai, Z.L.; Wu, G.; Zhu, W.Y. Amino acid metabolism in intestinal bacteria: Links between gut ecology and host health. *Front. Biosci.* **2011**, *16*, 1768–1786. [CrossRef] [PubMed]

58. North, M.K.; Dalle Zotte, A.; Hoffman, L. Composition of rabbit caecal microbiota and the effects of dietary quercetin supplementation and sex thereupon. *World Rabbit Sci.* **2019**, *27*, 185–198. [CrossRef]
59. Dao, M.C.; Everard, A.; Aron-Wisnewsky, J.; Sokolovska, N.; Prifti, E.; Verger, E.O.; Kayser, B.D.; Levenez, F.; Chilloux, J.; Hoyles, L.; et al. *Akkermansia muciniphila* and improved metabolic health during a dietary intervention in obesity: Relationship with gut microbiome richness and ecology. *Gut* **2016**, *65*, 426–436. [CrossRef]
60. Gómez-Gallego, C.; Pohl, S.; Salminen, S.; De Vos, W.M.; Kneifel, W. *Akkermansia muciniphila*: A novel functional microbe with probiotic properties. *Benef. Microbes* **2016**, *7*, 571–584. [CrossRef]
61. Zhai, Q.; Feng, S.; Arjan, N.; Chen, W. A next generation probiotic, *Akkermansia muciniphila*. *Crit. Rev. Food Sci. Nutr.* **2019**, *59*, 3227–3236. [CrossRef]
62. Dunfield, P.F.; Yuryev, A.; Senin, P.; Smirnova, A.V.; Stott, M.B.; Hou, S.; Ly, B.; Saw, J.H.; Zhou, Z.; Ren, Y.; et al. Methane oxidation by an extremely acidophilic bacterium of the phylum Verrucomicrobia. *Nature* **2007**, *450*, 879–882. [CrossRef]
63. Martinez-Garcia, M.; Brazel, D.M.; Swan, B.K.; Arnosti, C.; Chain, P.S.G.; Reitenga, K.G.; Xie, G.; Poulton, N.J.; Gomez, M.L.; Masland, D.E.D.; et al. Capturing single cell genomes of active polysaccharide degraders: An unexpected contribution of verrucomicrobia. *PLoS ONE* **2012**, *7*, e35314. [CrossRef]
64. Wertz, J.T.; Kim, E.; Breznak, J.A.; Schmidt, T.M.; Rodrigues, J.L. Genomic and physiological characterization of the *Verrucomicrobia* isolate *Diplosphaera colitermitum* gen. Nov., sp. nov., reveals microaerophily and nitrogen fixation genes. *Appl. Environ. Microbiol.* **2012**, *78*, 1544–1555. [CrossRef]
65. Liu, C.; Finegold, S.M.; Song, Y.; Lawson, P.A. Reclassification of *Clostridium coccoides, Ruminococcus hansenii, Ruminococcus hydrogenotrophicus, Ruminococcus luti, Ruminococcus productus* and Ruminococcus *schinkii* as *Blautia coccoides* gen. nov., comb. nov., *Blautia hansenii* comb. nov., *Blautia hydrogenotrophica* comb. nov., *Blautia luti* comb. nov., *Blautia producta* comb. nov., *Blautia schinkii* comb. nov. and description of *Blautia wexlerae* sp. nov., isolated from human faeces. *Int. J. Sys. Evol. Microbiol.* **2008**, *58*, 1896–1902.
66. Chen, J.; Zhang, H.; Wu, X.; Shang, S.; Yan, J.; Chen, Y.; Zhang, H.; Tang, X. Characterization of the gut microbiota in the golden takin (*Budorcas taxicolor bedfordi*). *AMB Express* **2017**, *7*, 1–10. [CrossRef] [PubMed]
67. Gulino, L.M.; Ouwerkerk, D.; Kang, A.Y.; Maguire, A.J.; Kienzle, M.; Klieve, A.V. Shedding light on the microbial community of the macropod foregut using 454-amplicon pyrosequencing. *PLoS ONE* **2013**, *8*, e61463. [CrossRef]
68. Tims, S.; Derom, C.; Jonkers, D.M.; Vlietinck, R.; Saris, W.H.; Kleerebezem, M.; De Vos, W.M.; Zoetendal, E.G. Microbiota conservation and BMI signatures in adult monozygotic twins. *ISME J.* **2013**, *7*, 707–717. [CrossRef] [PubMed]
69. Clark, A.; Mach, N. Exercise-induced stress behavior, gut-microbiota-brain axis and diet: A systematic review for athletes. *J. Int. Soc. Sports Nutr.* **2016**, *13*, 43. [CrossRef] [PubMed]
70. Gosalbes, M.J.; Durbán, A.; Pignatelli, M.; Abellan, J.J.; Jiménez-Hernández, N.; Pérez-Cobas, A.E.; Latorre, A.; Moya, A. Metatranscriptomic approach to analyze the functional human gut microbiota. *PLoS ONE* **2011**, *6*, e17447. [CrossRef]

Article

Effect of Thymol Addition and Withdrawal on Some Blood Parameters, Antioxidative Defence System and Fatty Acid Profile in Rabbit Muscle

Kristina Bacova [1], Karin Zitterl-Eglseer [2], Lubica Chrastinova [3], Andrea Laukova [1], Michaela Madarova [1], Sona Gancarcikova [4], Drahomira Sopkova [5], Zuzana Andrejcakova [5] and Iveta Placha [1,*]

1 Centre of Biosciences, Slovak Academy of Sciences, Institute of Animal Physiology, Soltesovej 4–6, 040 01 Kosice, Slovakia; bacovak@saske.sk (K.B.); laukova@saske.sk (A.L.); madarova@saske.sk (M.M.)

2 Institute of Animal Nutrition and Functional Plant Compounds, University of Veterinary Medicine Vienna, Veterinärplatz 1, A-1210 Wien, Austria; karin.zitterl@vetmeduni.ac.at

3 National Agricultural and Food Centre, Hlohovecka 2, 951 41 Nitra-Luzianky, Slovakia; lubica.chrastinova@nppc.sk

4 Laboratory of Gnotobiology, Department of Microbiology and Immunology, University of Veterinary Medicine and Pharmacy, Komenskeho 73, 041 81 Kosice, Slovakia; sona.gancarcikova@uvlf.sk

5 Department of Anatomy, Histology and Physiology, University of Veterinary Medicine and Pharmacy, Komenskeho 73, 041 81 Kosice, Slovakia; drahomira.sopkova@uvlf.sk (D.S.); zuzana.andrejcakova@uvlf.sk (Z.A.)

* Correspondence: placha@saske.sk; Tel.: +421-55-792-2969

Received: 19 May 2020; Accepted: 20 July 2020; Published: 22 July 2020

Simple Summary: So far, the study of the bioactivity of thymol, a major constituent of *Thymus vulgaris* L., in the animal organism has received little attention. Our study could give us answers to questions about whether thymol accumulates in the rabbit organism after its sustained administration and if it is also able to exhibit its beneficial properties for a longer period. Thymol in powder form at the concentration 250-mg/kg feed was added to the rabbit diet for 21 days and withdrawn for the next seven days. We confirmed that thymol was sufficiently absorbed from the gastrointestinal tract and was able to express its biological activity not only during application but, also, after withdrawal. Further studies are needed to clarify the biotransformation and bioavailability of thymol in the rabbit organism with respect to the specific features of rabbit digestion.

Abstract: Thymol concentrations in rabbit plasma, intestinal wall (IW) and faeces were detected, and the effects of thymol application and withdrawal on biochemical, antioxidant parameters and fatty acids (FA) in blood (B) and muscle (M) were studied. Forty-eight rabbits were divided into two experimental groups (control, C and with thymol 250-mg/kg feed, T). Thymol was administered for 21 days (TA) and withdrawn for seven days (TW). Thymol in plasma correlated with that in the IW (Spearman's correlation coefficient (r_s) = −1.000, $p = 0.0167$, TA) and was detected in faeces (TA and TW). In TA alkaline phosphatase ($p = 0.0183$), cholesterol ($p = 0.0228$), malondialdehyde ($p = 0.003$), glutathione peroxidase ($p = 0.0177$) in B and lactate dehydrogenase (M, $p = 0.0411$) decreased; monounsaturated FA ($p = 0.0104$) and α-linolenic acid ($p = 0.0227$) in M increased. In TW urea ($p = 0.0079$), docosapentaenoic acid ($p = 0.0069$) in M increased; linoleic acid ($p = 0.0070$), ∑ n−6 ($p = 0.0007$) in M and triglycerides decreased (B, $p = 0.0317$). In TA and TW, the total protein ($p = 0.0025$ and 0.0079), creatinine (B; $p = 0.0357$ and 0.0159) and oleic acid (M; $p = 0.0104$ and 0.0006) increased. Thymol was efficiently absorbed from the intestine and demonstrated its biological activity in blood and the muscles.

Keywords: rabbit; thymol; bioavailability; antioxidant

1. Introduction

The supplementation of human and animal diets with *Thymus vulgaris* (thyme) either as dried leaves or its essential oil has often demonstrated its beneficial properties. Since synthetic growth promoters were replaced with alternative herbal products in animal rearing, thyme has started receiving major attention. The most important bioactive compound contained in this plant is thymol, which exhibits antimicrobial, antioxidant, anticarcinogenic and anti-inflammatory activities [1–3]. According to some studies, thyme improves the performance parameters, but some other studies suggest that it has no effect. Some studies report that thyme is able to reduce levels of triglyceride and total cholesterol [4,5]. Yu et al. [6] found that thymol possesses a lipid-reducing function by altering hepatic triglyceride secretion.

The mode of action of herbs and plant extracts and the details about the accumulation of phenolic substances in animal tissues are not known or not completely understood [7,8]. In general, the bioavailability of dietary compounds depends on their digestive stability and the efficiency of their transepithelial passage [9]. Moreover, their influence on each other's intestinal absorption has to be taken into account in studies concerned with the bioavailability of essential oil compounds.

Thymol as one of the major constituents of thyme oil presents a wide range of functional possibilities in pharmacy and the food industry. Besides thymol, thyme contains high concentrations of monoterpene phenols like carvacrol, p-cymene, 1,8-cineole, linalool, borneol, camphor, β-caryophyllene, thymol methyl ether and carvacrol methyl ether, which could have influenced the thymol absorption [2,3].

Ocel'ová et al. [10] were the first who analysed the thymol concentrations in individual intestinal segments in hens. Sufficient absorption of thymol from the digestive tract and its transport by systemic circulation to tissues in broiler chickens after four weeks of thyme essential oil application were demonstrated by Ocel'ová et al. [7]. Placha et al. [2] pointed to the sparing effect of thymol against oxidative deterioration of the antioxidant defence system in poultry after sustained thyme oil dietary application at 0.05% (thymol content 248.97 mg/kg dry matter (DM)) and 0.1% (thymol content 460.22 mg/kg DM) concentrations.

Concerning the toxicity of thymol, the data are controversial. According to a recent report of the European Food Safety Authority (EFSA) [11], thymol, when administered by the oral route in a rat, mouse and guinea pig (lethal dose - LD 50, 0.98, 1.80 and 0.88 mg/kg, respectively), suggested a moderately acute toxicity. No significant alteration was observed when a thymol oil-water emulsion was administered at doses (15.39, 30.78 and 61.55 mg/kg, respectively) during 28 days [12]. Based on European Commission (EC) [13], no maximum residual limit for thymol in foodstuffs of animal origin is needed to establish when it is used as a veterinary medicinal product. These data show that further studies are required to establish the thymol appropriate concentration.

Rabbits have a unique digestive system that is represented by an original feature of rabbit feeding behaviour named caecotrophy. Caecotrophy means the excretion and immediate consumption of specific soft faeces termed "caecotrophs". This process is extremely important, because it improves feed utilization by maximizing the digestibility of nutrients [14]. Many scientific studies have tried to find the appropriate dose and form of plant extract application for improving animal health, but insights into the precise mechanism or mode of action of their components are lacking. As far as rabbits are concerned, to our knowledge, only one old study by Takada et al. [15] has been carried out regarding the metabolic outcome of thymol. Probably, the process of caecotrophy could minimize the loss of thymol bioaccessibility during the digestive processes.

The present study provides new insights for understanding the possible processes of absorption and deposition of thymol in the rabbit organism in connection with its protective role against oxidative stress. Based on our previous studies related to the thymol absorption, deposition and beneficial effects against oxidative stress in broiler chickens after a sustained administration of thyme oil, we decided to examine the thymol application into rabbit diets at the concentration 250-mg/kg feed.

2. Materials and Methods

2.1. Animals and Experimental Design

After weaning at 35 days of age, 48 rabbits of both sexes (meat line M9) were randomly divided into two experimental groups (control, C and with thymol addition, T), with six replicates in each (one replicate consisting of two cages, one cage/two animals). Initial live weight was 1006 ± 98 g in C and 1035 ± 107 g in T. All experimental wire-net cages (61 cm × 34 cm × 33 cm) were kept in rooms with automatic temperature control (22 ± 4 °C) and photoperiod (16 L:8 D). The rabbits could feed ad libitum and had free access to drinking water. The experiment lasted 28 days. The rabbits were fed with thymol addition for 21 days (TA), and for the next 7 days, the thymol was withdrawn (TW). Eight starved rabbits (6 male and 2 female) in each group were slaughtered in an experimental slaughterhouse at 56 (C and TA) or 63 (C and TW) days of age. Rabbits were stunned with electronarcosis (50 Hz, 0.3 A/rabbit for 5 s), immediately hung by the hind legs on the processing line and quickly bled by cutting the jugular veins and the carotid arteries.

2.2. Animals Care and Use

The trial was carried out at the experimental rabbit facility of the National Agricultural and Food Centre, Research Institute for Animal Production, Nitra, Slovakia. The protocol was approved by the Institutional Ethical Committee, and the State Veterinary and Food Office of the Slovak Republic approved the experimental protocol (4047/16-221).

2.3. Diets and Chemical Analyses

The basal diet (control, C) was formulated to satisfy growing rabbits' requirements [16] (Table 1) and was tested against the experimental diet (T) containing thymol (≥99.9%, Sigma Aldrich, St. Louis, MO, USA), which was added to the basal diet in white powder form at concentration 250-mg/kg feed. The diets were administered in the form of pellets with an average size of 3.5 mm. The feed was stored in darkness to protect against degradative processes and was analysed to determine the crude protein (CP), ash, ether extract, acid detergent fibre (ADF), starch and dry matter (DM) in the diets, while DM was also determined for the intestinal wall, muscle, liver and faeces according to the Association of Official Analytical Chemists (AOAC) methods [17]. Neutral detergent fibre (NDF) was analysed according to Van Soest et al. [18].

Table 1. Ingredients and chemical composition of rabbit diets.

Ingredients (%)		Chemical Composition (g/kg Feed)	
Dehydrated Lucerne meal	36.0	Dry matter (g/kg)	900.9
Dry malting sprouts	15.0	Organic compounds	831.8
Oats	13.0	Nitrogen free extract	444.3
Wheat bran	9.0	Neutral detergent fibre (NDF)	352.9
Barley	8.0	Acid detergent fibre (ADF)	208.1
Extracted sunflower meal	5.5	Crude fibre	177.8
Extracted rapeseed meal	5.5	Crude protein	176.6
Dried distiller grains with solubles	5.0	Cellulose	163.1
Premix [1]	1.7	Hemicellulose	144.8
Limestone	1.0	Starch	133.1
Sodium chloride	0.3	Ash	69.2
		Fat	33.1
		Metabolic energy, MJ/kg	9.9

[1] The vitamin-mineral premix provided per kg of complete diet: Retinyl acetate 5.16 mg, Cholecalciferol 0.03 mg, Tocopherol 0.03 mg, Thiamine 0.8 mg, Riboflavin 3.0 mg, Pyridoxin 2.0 mg, Cyanocobalamin 0.02 mg, Niacin 38 mg, Folic acid 0.6 mg, Calcium 1.8 mg, Iron 70 mg, Zinc 66 mg, Copper 15 and Selenium 0.25 mg.

2.4. Growth Performance and Health Status

Body weight (BW) and feed intake (FI) were recorded individually once a week. The average daily FI, average daily weight gain (WG) and feed conversion ratio (FCR) were calculated at the end of the trial (on 56 and 63 d of age). Data pertaining to any animal that died during the experiment were excluded from the calculations. Mortality was recorded daily throughout the experimental periods.

2.5. Thymol Antioxidant Capacity and Stability in Feed

The Trolox equivalent capacity (TEAC) was determined in thymol and in the experimental feed according with the method described by Karamać et al. [19] using the 2,2′-Azinobis-(3-Ethylbenzthiazolin-6-Sulfonic Acid (ABTS •+) decolorization assay. The results were expressed as mmol Trolox equivalents (TE) per g. Thymol evaporation in the feed was analysed every week during thymol application using High-Performance Liquid Chromatography HPLC according to the modified method of Pisarčíková et al. [20]. Samples were analysed in triplicate and were relatively stable (0 d—151.89, 7 d—134.75 and 14 d—128.30 µg/g DM, respectively).

2.6. Sampling

Blood samples for biochemical analyses were collected from the marginal ear vein (Vena auricularis) into dry nonheparinized Eppendorf tubes at experimental days 21 and 28 and were left to clot in a standing position for approximately 2 h to obtain the serum, and then, the serum was separated by centrifugation at 700× *g* for 15 min. Blood for analyses of antioxidant parameters was collected into heparinized tubes, and plasma was obtained after centrifugation at 1180× *g* for 15 min. Samples of serum, plasma, muscle (*Longissimus thoracis et lumborum*, LTL), small intestinal wall, liver and hard faeces (freshly voided, collected using nets mounted under the cages) were immediately frozen in liquid nitrogen and stored at −70 °C until analysis.

2.7. Thymol Analyses in Plasma, Small Intestinal Wall and Faeces

Detection of thymol in samples of plasma, intestinal walls and faeces was performed using headspace solid-phase microextraction followed by gas chromatography coupled with the mass spectrometry method, as described by Placha et al. [2]. Briefly, detection and quantification were carried out using a gas chromatography/mass spectrometry (GC/MS) (type HP 6890 GC) coupled with a 5972 quadrupole-mass selective detector (Agilent Technologies GmbH, Wilmington, DE, USA). Detection of thymol was confirmed by comparing its specific mass spectrum and retention time with those of the reference compound. Additionally, the Kovats index was calculated. Calibration curves were generated by plotting the peak area ratios of thymol to o-cresol used as the internal standard (Sigma-Aldrich, St. Louis, MO, USA) against the known thymol concentrations. The selective ion mode was used for the quantitative analysis of thymol. The mass fragments *m*/*z* 135 and *m*/*z* 150, as well as *m*/*z* 107 and *m*/*z* 108, were monitored as characteristic for thymol and o-cresol, respectively. Calibration curves were prepared from blank samples spiked directly with thymol (AppliChem, Darmstadt, Germany) in standard solutions with known concentrations. Each point on the calibration curve was analysed as a duplicate. The peak of thymol was detected around 19 min, and the o-cresol peak occurred around 10 min in all samples. Samples for the detection of thymol were prepared using the method described by Ocel'ová et al. [10]. Enzyme β–Glucuronidase Helix pomatia Type HP-2 (aqueous solution, ≥100,000 units/mL, Sigma-Aldrich, St Louis, MO, USA) was added to samples to cleave thymol from its glucuronide and sulphate to obtain the total amount of thymol in the plasma.

2.8. Biochemical and Antioxidant Parameters and Activity of Lactate Dehydrogenase in Blood and Tissues

Total proteins (TP; g/L), creatinine (µmol/L), urea (mmol/L), triglycerides (mmol/L), total cholesterol (mmol/L), alanine aminotransferase (ALT; µkat/L), aspartate aminotransferase (AST; µkat/L) and alkaline phosphatase (ALP; μkat/L) were analysed using a DIALAB commercial kit (Prague, Czech Republic) and an ELLIPSE analyser (AMS, Guidonia, Rome, Italy).

Activity of glutathione peroxidase (GPx, EC 1.11.1.9) in blood was measured by monitoring the oxidation of Nicotinamide Adenine Dinucleotide Phosphate (NADPH) at 340 nm in-line with Paglia and Valentine [21] using a commercial kit (Ransel, Randox, London, UK). Haemoglobin (Hb) content in blood was analysed using a commercial kit from Randox, UK. The samples of LTL muscle and liver for malondialdehyde (MDA) measurement and activity of lactate dehydrogenase (LDH, EC 1.1.1.27) were washed in buffered saline to remove excess blood and connective tissue. Samples for MDA analyses were homogenised with deionized distilled water and 50 μL of 7.2% butylated hydroxytoluene and for LDH activity in cold buffer (0.05-mol/L Tris-HCl buffer, pH 7.3). The homogenates were subsequently centrifuged at 13,680× *g* at 4 °C for 20 min. MDA concentrations in these tissues and plasma were measured using the modified fluorometric method of Jo and Ahn [22]. The enzyme activity of LDH was measured using a commercial diagnostic kit (Crumlin, Randox, UK) with an Alizé automatic biochemical analyser (Lisabio, Pouilly-en-Auxois, France) at 340 nm, as described by Andrejčáková et al. [23]. The protein concentration in the muscle and liver was quantified using the spectrophotometric method published by Bradford [24].

2.9. Fatty Acids in Muscle Tissue

The fatty acid (FA) composition in the muscle tissue was determined using the method of Ouhayoun et al. [25]. Fatty acid methyl esters (FAME) were prepared by means of alcoholises in an essential nonalcoholic solution and analysed using gas chromatography on GC 6890N (Agilent Technologies, Basel, Switzerland). Results were expressed as percentages of total fatty acids.

2.10. Statistical Analyses

Values of thymol, GPx, LDH, MDA and FA concentrations were tested for normal distribution with the Kolmogorov-Smirnov test. The Mann-Whitney U test was used for statistical analysis. Results were presented as the mean value ± standard deviation. Significant differences were considered at $p < 0.05$. Correlations of thymol concentrations between plasma and feed and plasma and intestinal wall were analysed using nonparametric Spearman's rank correlation and expressed as Spearman's correlation coefficient (r_s). Statistical analyses were performed using Graph Pad Prism 5.0. (GraphPad Software, San Diego, CA, USA).

3. Results

3.1. Growth Performance

All broiler rabbits in the present trial were in good health, and the growth performance was normal and was not affected by the addition of thymol. Five animals (three/control and two/experimental group) died during the whole experiment. If there were no differences in the given parameters, we do not include them in the table.

3.2. Thymol Antioxidative Capacity

Approximately equal TEAC value was found in thymol and in the feed of the experimental group (0.76 vs. 0.74 mmol TE/g).

3.3. Thymol in Feed, Plasma, Small Intestinal Wall and Faeces

Thymol content in feed amounted to 148.9 ± 16.7 μg/g DM. Thymol concentration in plasma was 0.05 ± 0.02 μg/L and, in intestinal wall, 0.04 ± 0.03 μg/g DM in TA, but, in TW, it was not detected. Thymol concentration in faeces in TA was 0.89 ± 0.45 μg/g DM, and, in TW, it was 0.08 ± 0.04 μg/g DM (Table 2). Thymol concentration in plasma significantly correlated with the thymol concentration in the intestinal wall ($r_s = -1.000$, $p = 0.0167$).

Table 2. Thymol content in plasma (μg/L), feed, intestinal wall and faeces (μg/g dry matter (DM)).

Substance	TA	TW
Feed	148.90 ± 16.65	-
Plasma	0.05 ± 0.02	ND
Intestinal wall	0.04 ± 0.03	ND
Faeces	0.89 ± 0.45	0.08 ± 0.04

ND—not detected, TA—thymol addition and TW—thymol withdrawal.

3.4. Biochemical Parameters in Blood

Thymol addition significantly decreased the levels of ALP ($p = 0.0183$) and cholesterol ($p = 0.0228$). Urea significantly increased ($p = 0.0079$), and triglycerides decreased ($p = 0.0317$) in TW. TA and TW had significantly increased TP ($p = 0.0025$ vs. $p = 0.0079$) and creatinine ($p = 0.0357$ vs. $p = 0.0159$) (Table 3).

Table 3. Effects of thymol on aspartate aminotransferase (AST, μkat/L), alanine aminotransferase (ALT, μkat/L), alkaline phosphatase (ALP, μkat/L), total proteins (TP, g/L), urea (mmol/L), creatinine (μmol/L), triglycerides (mmol/L) and cholesterol (mmol/L) in rabbit blood.

Parameter	TA			*p*-Value	TW			*p*-Value
	C	T	SD		C	T	SD	
AST (μkat/L)	0.45	0.34	0.12	0.1199	0.49	0.34	0.20	0.2948
ALT (μkat/L)	0.77	0.62	0.14	0.2677	0.76	0.69	0.32	1.0000
ALP (μkat/L)	2.00 ^{a}	1.52 ^{b}	0.42	0.0183	2.42	2.19	0.83	0.5476
TP (g/L)	36.20 ^{a}	61.86 ^{b}	13.88	0.0025	36.84 ^{a}	55.90 ^{b}	11.79	0.0079
Urea	5.18	5.31	0.50	1.0000	5.03 ^{a}	6.54 ^{b}	0.95	0.0079
Creatinine	79.67 ^{a}	115.40 ^{b}	19.55	0.0357	64.25 ^{a}	115.60 ^{b}	28.52	0.0159
Triglycerides	1.15	1.45	0.29	0.1508	1.53 ^{a}	0.84 ^{b}	0.49	0.0317
Cholesterol	2.20 ^{a}	1.17 ^{b}	0.80	0.0228	1.76	0.95	0.60	0.0556

^{a,b} Values within a row with different superscript letters differed significantly ($p < 0.05$). Data are presented as mean ± standard deviation (SD). C—control diet, TA—thymol addition and TW—thymol withdrawal.

3.5. Antioxidant Parameters in Blood and Tissues

MDA and GPx in blood ($p = 0.003$ and $p = 0.0177$) and LDH in muscle ($p = 0.0411$) significantly decreased in TA (Table 4).

Table 4. Effects of thymol on the antioxidant parameters and activity of lactate dehydrogenase (LDH) in rabbit blood, liver and muscle.

Parameter	TA			*p*-Value	TW			*p*-Value
	C	T	SD		C	T	SD	
Blood								
MDA (nmol/mL)	0.35 ^{a}	0.254 ^{b}	0.06	0.0030	0.32	0.30	0.04	0.1049
GPx (μkat/g Hb)	3.61 ^{a}	2.52 ^{b}	0.78	0.0177	2.61	2.32	0.71	0.8763
Liver								
MDA (nmol/g protein)	83.11	83.40	18.31	0.7984	83.81	68.69	21.51	0.1605
LDH (μkat/g protein)	41.56	41.62	6.50	0.6454	48.88	42.75	7.83	0.1304
Muscle								
MDA (nmol/g protein)	23.63	28.45	11.20	0.5737	18.34	23.45	5.82	0.0650
LDH (μkat/g protein)	78.74 ^{a}	35.89 ^{b}	36.47	0.0411	78.50	50.62	43.60	0.2026

^{a,b} Values within a row with different superscript letters differed significantly ($p < 0.05$). Data are presented as mean ± standard deviation (SD). C—control diet, TA—thymol addition, TW—thymol withdrawal, MDA—malondialdehyde and GPx—glutathione peroxidase.

3.6. Fatty Acids in Muscle

Concentrations of oleic acid (C 18:1 n-9) significantly increased in TA and TW ($p = 0.0104$ vs. 0.0006). MUFA and α-linolenic acid (C 18:3 n-3) significantly increased in TA ($p = 0.0104$ and $p = 0.0227$).

Docosapentaenoic acid (C 22:5 n-3) significantly increased ($p = 0.0069$), and linoleic acid (C 18:2 n-6) and ∑ n-6 significantly decreased in TW ($p = 0.0070$ and $p = 0.0007$; Table 5).

Table 5. Effects of thymol on fatty acids profile (% of total FA) in rabbit muscle.

Parameter	TA			*p*-Value	TW			*p*-Value
	C	T	SD		C	T	SD	
C 12:0	0.107	0.109	0.006	0.3428	0.105	0.108	0.004	0.1015
C 14:0	1.383	1.392	0.030	0.5950	1.357	1.386	0.039	0.1375
C 16:0	24.400	24.320	0.159	0.3823	24.390	24.580	0.222	0.1049
C 18:0	10.460	10.550	0.251	0.5049	10.750	10.620	0.210	0.2392
C 17:0	0.302	0.308	0.029	0.3713	0.312	0.286	0.037	0.4001
∑ SFA	36.650	36.680	0.344	1.0000	36.910	36.980	0.280	1.0000
C 18:1 n-7	4.945	4.975	0.109	0.6454	4.854	4.899	0.105	0.3439
C 18:1 n-9	30.030	34.470	3.754	0.0104	31.960	37.270	3.660	0.0006
C 20:1	0.597	0.589	0.091	0.8785	0.545	0.561	0.072	0.5054
∑ MUFA	35.600	40.020	3.730	0.0104	37.420	42.670	4.640	0.2345
C 18:2 n-6	52.59	50.73	6.435	0.5054	51.36	48.83	8.458	0.0070
C 20:4 n-6	1.765	1.978	0.244	0.1049	1.842	1.708	0.279	0.5054
∑ n-6	54.35	52.70	6.341	0.5054	53.20	50.53	4.756	0.0007
C 18:3 n-3	1.794	1.958	0.149	0.0227	1.793	1.885	0.191	0.2786
C 20:5 n-3	1.088	1.108	0.103	0.6350	1.096	0.935	0.173	0.1409
C 22:5 n-3	1.270	1.393	0.127	0.0513	1.276	1.381	0.081	0.0069
C 22:6 n-3	0.328	0.298	0.049	0.3431	0.341	0.310	0.046	0.2677
∑ n-3	4.479	4.755	0.236	0.0519	4.506	4.511	0.199	1.000
∑ n-6/∑ n-3	12.14	11.10	1.444	0.1949	11.86	11.19	1.223	0.3282

Data are presented as mean ± standard deviation (SD). C—control diet, TA—thymol addition, TW—thymol withdrawal, SFA—saturated fatty acids and MUFA—monounsaturated fatty acids.

4. Discussion

4.1. Growth Performance

The inclusion of thymol at the concentration used in our experiment did not show any significant effect on the animals' weight, weight gain and conversion ratio. Although thyme has been shown to improve the palatability and feed intake in growing rabbits, its beneficial effects on the live growth performance in rabbits has not yet been confirmed [26]. According to Erdelyi et al. [27], due to the specific digestive physiology of rabbits, essential oils or plant extracts have much lesser positive effects than in broilers or piglets. Windisch et al. [28] reported that the use of thyme in animal diets could be limited when applied in certain amounts, because it is highly aromatic. According to Gerencsér et al. [29], thyme leaves did not demonstrate any substantial effects on the growth performance or health status. This statement is in agreement with our study, in that the thymol did not affect the feed intake, and most animals remained in optimal condition.

4.2. Thymol in Plasma, Intestinal Wall and Faeces

We found a significant correlation between thymol levels in the small intestinal wall and plasma during the period of thymol addition to the feed. This result is in agreement with Placha et al. [2], who confirmed the efficient absorption of thymol from the digestive tract into the systemic circulation in broiler chickens. According to Ocel'ová [30], after the absorption of plant compounds from the intestine, they are metabolised and eliminated from the organism.

One of the most original features of rabbit-feeding behaviour is caecotrophy. The result of this process, in which soft faeces are swallowed and then stored intact in the fundus of the stomach, is that they undergo the same digestive processes as normal feed. Some parts of the initial food intake may be recycled even up to four times in this way [31]. The detection of thymol in faeces points to its excretion from the organism in an unmetabolised form (Table 2). This unmetabolised thymol could be recycled and again absorbed in the intestinal wall, where it is finally metabolised by the processes of biotransformation. The part of the metabolites in the intestinal wall can be transported back into the intestinal lumen by efflux transporters, converted to thymol (parental compound) and again reabsorbed by the enterocytes. Another molecule can be transported through the basolateral membrane of enterocytes directly to the blood circulation. All these processes play a crucial role in affecting the thymol concentration in blood [7].

Thymol sulphate and thymol glucuronide are the main metabolites of thymol biotransformation, and so far, little is known about their bioactivity or whether these compounds are only inactive forms [3]. Pisarčíková et al. [20] first detected thymol sulphate and thymol glucuronide in the liver and duodenal wall of broiler chickens after a sustained four-week administration of thyme essential oil, and according to Kohlert et al. [32] and Rubió et al. [9], thymol metabolites could be deconjugated to the parental compounds and, in this way, express their pharmacological properties. We have to bear in mind that, during biotransformation, plant compounds change their pharmacological properties, which usually differ from the properties of the parental compounds.

There are only a few studies concerning thymol distributions in animal tissues. Thymol conjugates have been detected in the plasma of humans and animals, and thymol after enzymatic cleavage was detected in the plasma of horses, chickens and pigs [1,10,32–35]. According to Placha et al. [2], erythrocytes can act as depots of polyphenols due to their ability to bind to the surface of red blood cells, and in this way, they can circulate in the organism. There is the question whether bound thymol is able to unbind from tissue depots if the amount of circulated thymol decreases. Since we detected thymol in rabbit faeces also after its withdrawal, this mechanism could be indicated (Table 2).

4.3. Biochemical Parameters in Blood

Despite the fact that plant bioactive compounds undergo fast biotransformation and elimination, Ocel'ová et al. [10] and Placha et al. [2] confirmed the accumulation of thymol in the breast muscles, kidneys and livers of broiler chickens after four weeks of thyme essential oil diet supplementation. Moreover, they found the highest concentration of thymol in the kidneys and the lowest in liver tissues, which could point to intensive metabolism in the liver and accumulation in the kidneys. Bardal et al. [36] showed that the distribution of plant compounds from systemic circulation to the tissues is restricted, since only free drugs in plasma are able to diffuse into the target tissue.

Gumus et al. [37] reported that specific polyphenols such as thymol may reduce plasma lipids by altering the hepatic triglyceride secretion, as well as inhibiting the activity of cholesterol-synthesizing enzyme 3-hydroxy-3-methylglutaryl-coenzyme A reductase. If some imbalance between the cellular free radical formation and the antioxidant defence system exists, excessive amounts of free radicals are produced, and cellular components such as the lipids are attacked. The administration of antioxidants such as thymol could combat oxidative stress by the scavenging of free radicals and, in this way, effectively reduce the serum levels of lipid parameters such as triglycerides and cholesterol [6]. Our study suggests that thymol in the rabbit organism possesses a lipid-reducing function by this mechanism (Table 3).

AST, ALT and ALP are enzymes that significantly reflect the liver function. These parameters in our study were in the range of the reference interval. We assumed that thymol in this concentration was able to express its antioxidant properties and positively affected these parameters, although only ALP was affected significantly. The values of urea and creatinine also remained within the normal ranges. This suggests that no hepatic or renal injuries occurred in this experiment. The slight increase in the urea and creatinine amounts could imply the effect of thymol and/or its metabolites on the kidney function, particularly on glomerular filtration. Ocel'ová [30] and Kohlert et al. [32] suggested the role of the kidneys in the metabolism and elimination of phenolic compounds. They assumed that thymol metabolites could be cleaved at the brush border of the renal tubule and reabsorbed as thymol back into the peritubular capillaries. As mentioned above, thymol can accumulate in the kidneys, and based on this finding, we can hypothesise that thymol concentrations in the kidneys in our experiment could lead during the metabolic processes to the formation of some toxic substances that could be responsible for altering the renal function (Table 3).

The nutritive value of the protein is determined not only by its amino acid composition but, also, by its digestibility [38]. The antioxidative and antimicrobial properties of thymol may beneficially affect the gastrointestinal microbiota ecosystem and, in this way, improve the nutrient utilisation. Moreover, phytogenics can stimulate the production of digestive enzymes and, also, beneficially affect the nutrient digestibility [39]. Ocel'ová [30] found that the duodenal wall in chickens was repeatedly exposed to thymol molecules after the four-week addition of thyme essential oil to their diet. Based on the significant correlation between thymol in the plasma and intestinal wall in our experiment, we assumed that the same processes were also present in the rabbit intestinal wall. Many studies have suggested that thymol possesses useful antioxidant properties that are responsible, among other things, for the renewal rate of mature enterocytes at the surface of the intestinal villi, which causes an increase in their absorption capacity [40]. Gasco et al. [41] reported that serum TP synthesis depends on the content of available protein in the diet. We assume that the increased TP in our groups with the thymol addition, as well as thymol withdrawal, was caused by extending the absorption surface of the intestinal wall, which probably evoked a better absorption of the proteins from the diet, and/or also by a higher production of the digestive enzymes (Table 3).

4.4. Antioxidant Parameters and LDH in Blood and Tissues

In the present study, the positive effect of thymol on the activity of GPx, which protects intracellular lipids against peroxidation, as well as on MDA levels as an indicator of lipid peroxidation, was expressed in plasma. This finding reflects the fact that thymol was efficiently absorbed from the digestive tract to the blood, where it manifested its antioxidant properties. The antioxidant properties of thymol were also demonstrated in muscles during the thymol addition, as well as after its withdrawal. Oxidative stress is associated with damage in a wide range of macromolecular species and derangements of the cellular metabolism, which permeabilise the cellular or organelle membranes and cause apoptosis [42]. LDH is an intracellular enzyme, and its increased level is an indicator of cellular damage. We detected decreased LDH levels, which points to the sparing effect of thymol on cellular metabolism (Table 4).

4.5. Fatty Acids in Muscle

Rabbits need adequate amounts of essential FA, which are primarily represented by linoleic (C18:2, n-6) and α-linolenic acids (C18:3, n-3) in their diet. The stability of these essential FA is very low, and for example, linolenic acid is oxidized ten times more rapidly than linoleic acid [43]. Polyunsaturated FA are easily oxidized because of the unstable double bonds between their carbon atoms and are rapidly broken down into short-chain compounds, which is undesirable from the nutritional aspect. The lipoperoxidation of the cell structure is a consequence of oxidative stress. These biological reactions are initiated by the production of ROS, which remove protons from FA [44]. Dabbou et al. [45] demonstrated that bilberry pomace, which is characterised by high antioxidant activity due to high contents of phenols, effectively prevented the oxidation of unsaturated lipids in the muscles of rabbits. An increase in unsaturated FA and, mainly, linolenic acid were observed in our study after the addition of thymol, which probably blocked the oxidation of the lipids with its strong antioxidative properties. After thymol withdrawal, linoleic acid, as well as $\sum$n-6, decreased, which could be the reason for the insufficient effect of thymol in connection with its low deposition in the muscles (Table 5). Moreover, as was mentioned above, linoleic acid is more rapidly oxidized than other polyunsaturated FA.

5. Conclusions

In conclusion, after its sustained application, thymol is able to accumulate and circulate in the rabbit organism, particularly due to the rabbits' specific digestive characteristic, i.e., caecotrophy. The investigated parameters were affected not only during the thymol addition to the feed but, also, after its withdrawal, which points to thymol accumulation in the organism. The thymol concentration used in this experiment clearly demonstrated its antioxidant properties and inhibited the lipid peroxidation and formation of oxidative deterioration compounds and may become an important antioxidant food supplement. However, further studies should be conducted to confirm the thymol distribution and deposition within the rabbit organism. Since there is a lack of literature describing the absorption and distribution of thymol in rabbit tissues in connection with its beneficial effects on animal health, to our knowledge, this is the first study attempting to explain these processes in the rabbit organism.

Author Contributions: Conceptualisation: I.P.; methodology: I.P. and K.B.; validation: I.P.; formal analysis: I.P., K.Z.-E., K.B., M.M., S.G., D.S. and Z.A.; investigation: I.P. and K.B.; resources: I.P., L.C., K.Z.-E., S.G. and D.S.; data curation: I.P., L.C. and A.L.; writing—original draft preparation: I.P. and K.B.; writing—review and editing: K.Z.-E.; visualisation: I.P. and K.B.; project administration: I.P. and funding acquisition: I.P. All authors have read and agreed to the published version of the manuscript.

Funding: This research was funded by the Scientific Grant Agency of the Ministry for Education, the Science, Research and Sport of the Slovak Republic and the Slovak Academy of Sciences (VEGA 2/0069/17, 2/0009/20 and 1/0204/20), as well as the Austrian Federal Ministry for Science, Research and Economics, OeAD, Ernst Mach Grant Action Austria-Slovakia.

Acknowledgments: The authors gratefully acknowledge the technical support provided by L. Ondruska, V. Parkanyi, R. Jurcik and J. Pecho from the National Agricultural and Food Centre, Research Institute for Animal Production, Nitra. The authors also thank Andrew Billingham for improving the written English of the manuscript.

Conflicts of Interest: The authors declare no conflicts of interest. The funders had no role in the design of the study; in the collection, analyses or interpretation of data; in the writing of the manuscript or in the decision to publish the results.

References

1. Haselmeyer, A.; Zentek, J.; Chizzola, R. Effects of thyme as a feed additive in broiler chickens on thymol in gut contents, blood plasma, liver and muscle. *J. Sci. Food Agric.* **2015**, *95*, 504–508. [CrossRef] [PubMed]
2. Placha, I.; Ocelova, V.; Chizzola, R.; Battelli, G.; Gai, F.; Bacova, K.; Faix, S. Effect of thymol on the broiler chicken antioxidative defence system after sustained dietary thyme oil application. *Br. Poult. Sci.* **2019**, *60*, 589–596. [CrossRef] [PubMed]
3. Salehi, B.; Mishra, A.P.; Shukla, I.; Sharifi-Rad, M.; Contreras, M.D.M.; Segura-Carretero, A.; Fathi, H.; Nasrabadi, N.N.; Kobarfard, F.; Sharifi-Rad, J. Thymol, thyme, and other plant sources: Health and potential uses. *Phytother. Res.* **2018**, *32*, 1688–1706. [CrossRef] [PubMed]
4. Mehdipour, Z.; Afsharmanesh, M.; Sami, M. Effects of supplemental thyme extract (Thymus vulgaris L.) on growth performance, intestinal microbial populations, and meat quality in Japanese quails. *Comp. Clin. Pathol.* **2014**, *23*, 1503–1508. [CrossRef]
5. Popović, S.; Puvača, N.; Kostadinović, L.; Džinić, N.; Bošnjak, J.; Vasiljević, M.; Djuragic, O. Effects of dietary essential oils on productive performance, blood lipid profile, enzyme activity and immunological response of broiler chickens. *Eur. Poult. Sci.* **2016**, 80. [CrossRef]
6. Yu, Y.M.; Chao, T.Y.; Chang, W.C.; Chang, M.J.; Lee, M.F. Thymol reduces oxidative stress, aortic intimal thickening, and inflammation-related gene expression in hyperlipidemic rabbits. *J. Food Drug Anal.* **2016**, *24*, 556–563. [CrossRef]
7. Ocel'ová, V.; Chizzola, R.; Battelli, G.; Pisarcikova, J.; Faix, S.; Gai, F.; Placha, I. Thymol in the intestinal tract of broiler chickens after sustained administration of thyme essential oil in feed. *J. Anim. Physiol. Anim. Nutr.* **2018**, *103*, 204–209. [CrossRef]
8. Botsoglou, N.A.; Christaki, E.; Fletouris, D.J.; Florou-Paneri, P.; Spais, A.B. The effect of dietary oregano essential oil on lipid oxidation in raw and cooked chicken during refrigerated storage. *Meat Sci.* **2002**, *62*, 259–265. [CrossRef]
9. Rubió, L.; Macià, A.; Castell-Auví, A.; Pinent, M.; Blay, M.T.; Ardévol, A.; Romero, M.P.; Motilva, M.J. Effect of the co-occurring olive oil and thyme extracts on the phenolic bioaccessibility and bioavailability assessed by in vitro digestion and cell models. *Food Chem.* **2014**, *149*, 277–284. [CrossRef]

10. Ocel'ová, V.; Chizzola, R.; Pisarčíková, J.; Novak, J.; Ivanišinová, O.; Faix, Š. Effect of thyme essential oil supplementation on thymol content in blood plasma, liver, kidney and muscle in broiler chickens. *Nat. Prod. Commun.* **2016**, *11*, 1545–1550. [CrossRef]
11. EFSA (European Food Safety Authority). Conclusion on the peer review of the pesticide risk assessment of the active substance thymol. *EFSA J.* **2012**, *10*, 1–43.
12. Gad, M.M.E. Acute and Repeated-Doses (28 Days) Toxicity of Thymol Formulation in Male Albino Rats. *Aust. J. Basic Appl. Sci.* **2012**, *7*, 915–922.
13. EC (European Commission). Commission Regulation (EU) No 37/2010 of 22 December 2009 on pharmacologically active substances and their classification regarding maximum residue limits in foodstuffs of animal origin. *Off. J. Eur. Union* **2010**, *L15*, 1–72.
14. Irlbeck, N.A. How to feed the rabbit (Oryctolagus cuniculus) gastrointestinal tract. *J. Anim. Sci.* **2001**, *79*, E343–E346. [CrossRef]
15. Takada, M.; Agata, I.; Sakamoto, M.; Yagi, N.; Hayashi, N. On the metabolic detoxication of thymol in rabbit and man. *J. Toxicol. Sci.* **1979**, *4*, 341–350. [CrossRef]
16. de Blas, C.; Mateos, G.G. Feed formulation. In *Nutrition of the Rabbit*, 2nd ed.; de Blas, C., Wiseman, J., Eds.; CAB International: Wallingford, UK, 2010; pp. 222–232.
17. AOAC (Association of official analytical chemists). *International Official Methods of Analysis*, 17th ed.; AOAC: Arlington, VA, USA, 2000.
18. Van Soest, P.J.; Robertson, J.B.; Lewis, B. A Methods for dietary fiber, neutral detergent fiber, and nonstarch polysaccharides in relation to animal nutrition. *J. Dairy Sci.* **1991**, *74*, 3583–3597. [CrossRef]
19. Karamać, M.; Gai, F.; Peiretti, P.G. Effect of the growth stage of false flax (Camelina sativa L.) on the phenolic compound content and antioxidant potential of the aerial part of the plant. *Polish J. Food Nutr. Sci.* **2020**, *70*, 189–198. [CrossRef]
20. Pisarčíková, J.; Oceľová, V.; Faix, Š.; Plachá, I.; Calderón, A.I. Identification and quantification of thymol metabolites in plasma, liver and duodenal wall of broiler chickens using UHPLC-ESI-QTOF-MS. *Biomed. Chromatogr.* **2017**, *31*, 1–12. [CrossRef]
21. Paglia, D.E.; Valentine, W.N. Studies on the quantitative and qualitative characterization of erythrocyte glutathione peroxidase. *J. Lab. Clin. Med.* **1967**, *70*, 158–169.
22. Jo, C.; Ahn, D.U. Fluorometric analysis of 2-thiobarbituric acid reactive substances in turkey. *Poult. Sci.* **1998**, *77*, 475–480. [CrossRef]
23. Andrejčáková, Z.; Sopková, D.; Vlčková, R.; Kulichová, L.; Gancarčíková, S.; Almášiová, V.; Holovská, K.; Petrilla, V.; Krešáková, L. Synbiotics suppress the release of lactate dehydrogenase, promote non-specific immunity and integrity of jejunum mucosa in piglets. *Anim. Sci. J.* **2016**, *87*, 1157–1166. [CrossRef] [PubMed]
24. Bradford, M. A rapid and sensitive method for the quantitation of microgram quantities of protein utilising the principle of protein-dye binding. *Anal. Biochem.* **1976**, *72*, 248–254. [CrossRef]
25. Ouhayoun, J. Rabbit meat characteristics and qualitative variability. *Cuni-Sciences* **1992**, *7*, 1–15.
26. Fekete, S.; Lebas, F. Effect of a natural flavour (Thyme extract) on the spontaneous feed ingestion, digestion coefficients and fattening parameters. *Magy. Allatorv. Lapja* **1983**, *38*, 121–125.
27. Erdelyi, M.; Matics, Z.; Gerencsér, Z.; Princz, Z.; Szendrő, Z.; Mézes, M. Study of the effect of rosemary (Rosmarinus officinalis) and garlic (Allium sativum) essential oils on the performance of rabbit. In *Nutrition and Digestive Physiology, Proceedings of the 9th World Rabbit Congress, Verona, Italy, 10–13 June 2008*; Xicato, G., Trocino, A., Lukefahr, S.D., Eds.; World Rabbit Science Association: Castanet-Tolosan, France, 2008.
28. Windisch, W.; Schedle, K.; Plitzner, C.; Kroismayr, A. Use of phytogenic products as feed additives for swine and poultry. *J. Anim. Sci.* **2008**, *86*, E140–E148. [CrossRef] [PubMed]
29. Gerencsér, Z.; Szendro, Z.; Matics, Z.; Radnai, I.; Kovács, M.; Nagy, I.; Cullere, M.; Dal Bosco, A.; Dalle Zotte, A. Effect of dietary supplementation of spirulina (Arthrospira platensis) and thyme (Thymus vulgaris) on apparent digestibility and productive performance of growing rabbits. *World Rabbit Sci.* **2014**, *22*, 1–9. [CrossRef]
30. Oceľová, V. Plant Additives in Relation to the Animal Gastrointestinal Tract and Metabolism of Their Main Compounds. Ph.D Thesis, Institute of Animal Physiology, Slovak Academy of Sciences, University of Veterinary Medicine and Pharmacy in Kosice, Košice, Slovakia, 2017.
31. Lebas, F.; Gidene, T. Feeding Behaviour in Rabbits. In Proceedings of the III International Rabbit Production Symposium, Villareal, Portugal, 2 November 2005.

32. Kohlert, C.; Schindler, G.; März, R.W.; Abel, G.; Brinkhaus, B.; Derendorf, H.; Gräfe, E.U.; Veit, M. Systemic availability and pharmacokinetics of thymol in humans. *J. Clin. Pharmacol.* **2002**, *42*, 731–737. [CrossRef]
33. Rubió, L.; Serra, A.; Macià, A.; Borràs, X.; Romero, M.P.; Motilva, M.J. Validation of determination of plasma metabolites derived from thyme bioactive compounds by improved liquid chromatography coupled to tandem mass spectrometry. *J. Chromatogr. B* **2012**, *905*, 75–84. [CrossRef]
34. Van den Hoven, R.; Zappe, H.; Zitterl-Eglseer, K.; Jugl, M.; Franz, C. Study of the effect of Bronchipret on the lung function of five Austrian saddle horses suffering recurrent airway obstruction (heaves). *Vet. Rec.* **2003**, *152*, 555–557. [CrossRef]
35. Zitterl-Eglseer, K.; Wetscherek, W.; Stoni, A.; Kroismayr, A.; Windisch, W. Bioavailability of essential oils of a phytobiotic feed additive and impact of performance and nutrient digestibility in weaned piglets. *Bodenkult.-J. Land Manag. Food Environ.* **2008**, *59*, 121–129.
36. Bardal, S.K.; Waechter, J.E.; Martin, D.S. *Applied Pharmacology*; Elsevier Saunders: Louis, MO, USA, 2011; pp. 17–34.
37. Gumus, R.; Ercan, N.; Imik, H. The effect of thyme essential oil (Thymus Vulgaris) added to quail diets on performance, some blood parameters, and the antioxidative metabolism of the serum and liver tissues. *Rev. Bras. Cienc. Avic.* **2017**, *19*, 297–304. [CrossRef]
38. Villamide, M.J.; Nicodemus, N.; Fraga, M.J.; Carabaño, R. Protein Digestion. In *Nutrition of the Rabbit*, 2nd ed.; de Blas, C., Wiseman, J., Eds.; CAB International: Wallingford, UK, 2010; pp. 222–232.
39. Dalle Zotte, A.; Sartori, A.; Bohatir, P.; Rémignon, H.; Ricci, R. Effect of dietary supplementation of Spirulina (Arthrospira platensis) and Thyme (Thymus vulgaris) on growth performance, apparent digestibility and health status of companion dwarf rabbits. *Livest. Sci.* **2013**, *152*, 182–191. [CrossRef]
40. Bozkurt, M.; Küçükyilmaz, K.; Çatli, A.U.; Özyildiz, Z.; Çinar, M.; Çabuk, M.; Çoven, F. Influences of an essential oil mixture supplementation to corn versus wheat-based practical diets on growth, organ size, intestinal morphology and immune response of male and female broilers. *Ital. J. Anim. Sci.* **2012**, *11*, e54. [CrossRef]
41. Gasco, L.; Dabbou, S.; Trocino, A.; Xiccato, G.; Capucchio, M.T.; Biasato, I.; Dezzutto, D.; Birolo, M.; Meneguz, M.; Schiavone, A.; et al. Effect of dietary supplementation with insect fats on growth performance, digestive efficiency and health of rabbits. *J. Anim. Sci. Biotechnol.* **2019**, *10*, 4. [CrossRef]
42. Jayakumar, S.; Madankumar, A.; Asokkumar, S.; Raghunandhakumar, S.; Gokula Dhas, K.; Kamaraj, S.; Divya, M.G.; Devaki, T. Potential preventive effect of carvacrol against diethylnitrosamine-induced hepatocellular carcinoma in rats. *Mol. Cell. Biochem.* **2012**, *360*, 51–60. [CrossRef]
43. Xiccato, G. Fat digestion. In *Nutrition of the Rabbit*, 2nd ed.; de Blas, C., Wiseman, J., Eds.; CAB International: Wallingford, UK, 2010; pp. 222–232.
44. Niki, E.; Yoshida, Y.; Saito, Y.; Noguchi, N. Lipid peroxidation: Mechanisms, inhibition, and biological effects. *Biochem. Biophys. Res. Commun.* **2005**, *338*, 668–676. [CrossRef]
45. Dabbou, S.; Renna, M.; Lussiana, C.; Gai, F.; Rotolo, L.; Kovitvadhi, A.; Brugiapaglia, A.; Helal, A.N.; Schiavone, A.; Zoccarato, I.; et al. Bilberry pomace in growing rabbit diets: Effects on quality traits of hind leg meat. *Ital. J. Anim. Sci.* **2017**, *16*, 371–379. [CrossRef]

Article

Comparison of Rabbit, Kitten and Mammal Milk Replacer Efficiencies in Early Weaning Rabbits

Panthiphaporn Chankuang [1], Achira Linlawan [1], Kawisara Junda [1], Chittikan Kuditthalerd [1], Tuksaorn Suwanprateep [1], Attawit Kovitvadhi [2,*], Pipatpong Chundang [2], Pornchai Sanyathitiseree [3] and Chaowaphan Yinharnmingmongkol [4]

1. Faculty of Veterinary Medicine, Kasetsart University, Bangkok 10900, Thailand; panthiphaporn.c@ku.th (P.C.); achira.l@ku.th (A.L.); kawisara.ju@ku.th (K.J.); chittikan.k@ku.th (C.K.); tuksaorn.s@ku.th (T.S.)
2. Department of Physiology, Faculty of Veterinary Medicine, Kasetsart University, Bangkok 10900, Thailand; pichandang@gmail.com
3. Department of Large Animal and Wildlife Clinical Science, Faculty of Veterinary Medicine, Kasetsart University, Nakorn Pathom 73140, Thailand; fvetpos@ku.ac.th
4. Animal Space Pet Hospital, Bangkok 10170, Thailand; taaoy@hotmail.com

* Correspondence: fvetawk@ku.ac.th; Tel.: +66-89-2022-677

Received: 19 May 2020; Accepted: 20 June 2020; Published: 23 June 2020

Simple Summary: A milk replacer must be given as the main diet to young rabbits that are separated from their mothers before they reach weaning age (31–35 days). This procedure, which is a rescue protocol, allows them to survive. Moreover, the early separation of young rabbits before weaning prevents negative consequences in lactating rabbits, which is beneficial to pet rabbit producers. Kitten (KMR®, Pet-Ag Inc., Hampshire, IL, USA: KMR) or mammal (Zoologic® Milk matrix 30/52, Pet-Ag Inc., Hampshire, IL, USA: MMR) milk replacers have generally been suggested for use in rabbits; however, rabbit milk has a unique composition. Therefore, a rabbit milk replacer (RMR) was formulated in this study for comparison with these commercial products. Early weaned rabbits at 18 days of age were fed daily with RMR, KMR or MMR until 36 days after birth, while a commercial pelleted diet and water were provided at an amount exceeding the normal intake. The results indicated that it is possible to use RMR as a milk replacer for rabbits without serious adverse consequences. However, the RMR group presented a lower trend in nutrient digestibility than the other groups, although there was no statistical significant difference. Therefore, prebiotics and/or probiotics should be added to RMR formulations to improve this parameter.

Abstract: Early weaned rabbits should be fed using a milk replacer in order to survive. Therefore, a rabbit milk replacer (RMR) was developed and compared with a kitten milk replacer (KMR®: KMR) and a mammal milk replacer (Zoologic® Milk matrix 30/52: MMR). Thirty-six native crossbred rabbits aged 18 days were divided into three experimental groups (six replicates/group, two rabbits/replicate), fed RMR, KMR or MMR daily until they were 36 days old and euthanized at 38 days, while a complete pelleted diet and water were provided *ad libitum*. No statistically significant differences were observed in growth performance parameters, water intake, faecal weight, nutrient digestibility, internal organ weight, caecal pH, caecal cellulose activity, number of faecal pellets and amount of crude protein intake ($p > 0.05$). Caecal amylase activity in the KMR group and caecal protease activity in the RMR group were higher than in the MMR group ($p < 0.05$). The villus height and crypt depth of the MMR group were greater than in the RMR and KMR group ($p < 0.05$). In conclusion, it is possible to feed RMR to early weaning rabbits without serious adverse effects. However, probiotics and/or prebiotics should be supplemented in milk replacers and their benefits studied.

Keywords: digestibility; enzyme activity; gut histology; milk replacer; rabbit

1. Introduction

The size of the pet market has increased sharply in recent years and was estimated to be around 131.7 billion US dollars in 2016 [1]. Moreover, the compounded annual growth rate of the global pet care market was forecast to be 4.9% between 2018 and 2025 due to changes in the new generation's lifestyle, such as living alone or child-free marriage [1]. Nevertheless, humans still need interactions with living things, which add social, medical, emotional and physical benefits to their lives [2]. Companion animals are one solution that can offer these benefits [2]. Although rabbits are not as popular as dogs and cats, they occupy third place among companion animals because they are clean, quiet, non-harmful and require little space [3].

Generally, rabbits are weaned at 31–35 days of age by rabbit producers in Thailand and other countries [4,5] and begin to consume pelleted diets around 18 days of age [6]. Milk replacer has been suggested as a means of feeding early weaned rabbits and orphaned rabbits and of solving the problem of female rabbits that do not produce milk, this being rabbits' major nutrient source for survival [7] as well as preventing gastrointestinal disease [8]. The early separation of young rabbits from their mother can prevent a negative energy balance due to lactation, which supports a higher production yield and reduces disease transmission from the mother to young rabbits, of benefit for pet rabbit producers [4,9]. In addition, the smallest rabbit breed (Netherland Dwarf) can produce around 4–6 kittens per litter; however, the milk yield of this breed is not sufficient to support their kittens, leading to a high mortality rate among young rabbits [5]. Furthermore, milk replacer can be used to rescue unweaned wild rabbits [8]. Therefore, the use of a milk replacer provides one solution to these problems.

Commercial rabbit milk replacer remains lacking or unavailable in some countries. For this reason, kitten milk replacer (KMR) has been suggested as a substitute [7]. However, although kitten milk replacer can be used in early weaning rabbits, its growth performance has been found to be inferior to rabbit milk for rabbits [8]. In addition, two milk replacer formulas using mixtures of kitten milk replacers—Fox Valley Ultraboost and/or Fox Valley 32/40 (Fox Valley Animal Nutrition, INC., Lakemoor, IL, USA)—have been used to rescue young desert cottontail (*Sylvilagus audubonii*) and eastern cottontail rabbits (*S. floridanus*), for which the mortality rate was 26–59% [8]. The high concentration of nutrients (fat, protein and energy), the near absence of lactose, the high proportion of medium-chain saturated fatty acids with bacteriostatic properties (C8:0 and C10:0) and the short milking period required constitute the unique characteristics of rabbit milk and feeding behaviour [6,10]. Therefore, milk replacer for rabbits should be formulated respecting the properties of real rabbit milk [10]. Moreover, cost-effectiveness is another problem for rabbit producers, owners and wildlife rescue center [8]. Therefore, this study aimed to compare the efficiency of a developed rabbit milk replacer with two commercial products (kitten and mammal milk replacers) based on the growth performance and health status of early weaning rabbits (18 days old).

2. Materials and Methods

2.1. Ethics Statement

This study was conducted following standard guidelines at the animal experimental unit, Faculty of Veterinary Medicine (Kasetsart University, Bangkok, Thailand) and was approved by the Institutional Animal Care and Use Committee of Kasetsart University, Bangkok, Thailand (ACKU62-VET-037).

2.2. Animals, Diets, Milk Replacer Preparation and Experimental Design

Thirty-six 18-day-old native crossbreed rabbits with initial body weights of 134 ± 6.31 g/head (mean ± standard deviation) were taken from a local rabbit farm (Saha farm, Kanchanaburi, Thailand). Rabbits were randomly separated into three experimental groups with equal numbers of each sex (six replicates per group and two rabbits per replicate) containing: (1) rabbits fed rabbit milk replacer, which was formulated in this study (RMR); (2) rabbits fed kitten milk replacer (KMR®; Pet-Ag Inc., Hampshire, IL, USA); and (3) rabbits fed mammal milk replacer (Zoologic® Milk matrix 30/52; Pet-Ag

Inc., Hampshire, IL, USA; MMR). Rabbits were placed in a stainless cage (35 cm × 35 cm × 35 cm) with controlled room temperature, light and humidity at 20 ± 2 °C, 16L:8D and 75 ± 10%, respectively. The experiment was conducted for 20 days until the rabbits reached 38 days of age. At the end of the experiments, one rabbit per replicate was euthanized by intraperitoneal injection with pentobarbital sodium at 100 mg/kg (Nembutal, Ceva corporate, France) [11] and the samples were collected for further analysis, while another rabbit of each replicate was returned to the farm and reared until it reached 60 days of age.

The formulation of the RMR, including the chemical composition of the milk replacer, complete pelleted diet and rabbit milk, is illustrated in Table 1. Every day, rabbits were fed 10 mL at 38 °C of a freshly prepared mixture of milk replacer powder and clean water using a sterile syringe at 7:00 until they reached weaning age (36 days), while clean water and complete commercial rabbit diets (Lee Feed Mill, Publ. Co., Ltd., Phetchaburi, Thailand) were provided *ad libitum* throughout the experiment. The KMR and MMR powders were diluted with warmed water at a 7:13 ratio and homogenized. The RMR consisted of two parts: a hydrogenated palm fat part and a mixed dried powder part, containing all ingredients except hydrogenated palm fat and polyoxyethylene (80) sorbitan monooleate. For RMR preparation, hydrogenated palm fat was heated in an 800 W microwave for 60 s, changing from solid to liquid form as a result. The dried mixed powder part was then mixed with warm water. Subsequently, the two parts were homogenized and polyoxyethylene (80) sorbitan monooleate was added as an emulsifier. The ratio of mixed dried powder to hydrogenated palm fat to warm water was 24.5:17.5:78. The dilution ratio was selected based on equal dry matter content between the milk replacers and the solubility of the milk mixture.

Table 1. Ingredients and chemical components of different milk replacers, rabbit milk and diet.

Items	Artificial Milk Replacers			Rabbit Diet [a]	Rabbit Milk [b]
	RMR	KMR	MMR		
Ingredients (%)					
Sodium caseinate	43.8	-	-	-	-
Skimmed milk powder	7.73	-	-	-	-
Hydrogenated palm fat	41.5	-	-	-	-
Monodicalcium phosphate	3.13	-	-	-	-
Limestone	2.30	-	-	-	-
Salt	0.94	-	-	-	-
Premix [c]	0.50	-	-	-	-
Polyoxyethylene (80) sorbitan monooleate	0.10	-	-	-	-
Chemical composition					
Dry matter (%FM)	95.8	96.2	95.2	90.7	29.8
Crude ash (%DM)	9.77	6.66	8.27	5.69	7.38
Crude protein (%DM)	41.4	48.5	30.1	16.1	41.3
Ether extract (%DM)	43.8	22.9	52.5	2.42	43.3
Crude fiber (%DM)	ND	ND	ND	25.2	ND
Nitrogen free extract (%DM) [d]	5.03	22.0	9.13	50.6	8.05
Metabolizable energy content (kcal/100g DM) [e]	535	441	584	254	540

RMR = Rabbit milk replacer which was performed in this study, KMR = Kitten milk replacer (KMR®, Pet-Ag Inc., Hampshire, IL, USA), MMR = Mammal milk replacer (Zoologic® Milk matrix 30/52, Pet-Ag Inc., Hampshire, IL, USA), FM = Fresh matter, DM = Dry matter, ND = Not detect; [a] A commercial pelleted diet for rabbits (Lee Feed Mill, Publ. Co., Ltd., Phetchaburi, Thailand); [b] Chemical composition of rabbit milk [10]; [c] Vitamin and mineral premix (Topmix-B111, Top Feed Mills Co., Ltd., Pathumthani, Thailand) were supplied per kilogram of diets at 4,800,000 IU of vitamin A; 1,200,000 IU of vitamin D3; 6000 IU of vitamin E; 600 mg of vitamin K; 600 mg of vitamin B1; 2200 mg of vitamin B2; 10,000 mg of vitamin B3; 800 mg of vitamin B6; 4 mg of vitamin B12; 48 mg of biotin; 4800 mg of Calcium pantothenate acid; 200 mg of folic acid; 24,000 mg of Zn, 16,000 mg of Fe; 32,000 mg of Mn; 32,000 mg of Cu; 200 mg of I; 40 mg of Se; 40 mg of Co; [d] Calculation [12]; [e] Calculation based on Atwater system [13].

2.3. *Perfomance, Digestibility and Faecal Evaluation*

The animals were weighed at 18, 24, 30 and 36 days of age, whereas average daily feed intake (ADFI), average daily weight gain (ADG), feed conversion ratio (FCR), water intake and weight of faeces output were evaluated at 19–24, 25–30 and 31–36 days of age. The apparent digestibility of dry matter, organic matter, ether extract and crude protein was conducted at 23–27 and 31–35 days of age and contained six replicates/groups. The procedures for feeding, faecal collection, chemical analysis and calculation were in accordance with [14]. Briefly, feed intake was measured during the period of the digestibility trial. Faeces were removed from the cage at 9:00 on the first day of the digestibility trial. Subsequently, all faeces on a net under the cage were collected at 9:00 for four days. The faeces were weighted immediately after collection, put in a sterile plastic bag and kept at −20 °C for further chemical composition analysis following the procedure of [14]. Another study was conducted, where the amount of daily faecal pellet excretion was measured by counting the dried faecal pellets between 19 and 36 days old from photos.

2.4. Internal Organs, Gut Histology and Caecal pH

The internal organ weight and the body weight of the euthanized rabbits were determined. The duodenal part of the small intestine was fixed in 10% buffered formalin for further villus morphometric evaluation. Briefly, small pieces of middle duodenum after fixation were processed, embedded in paraffin, sectioned at 7-μm thicknesses by means of a rotary microtome (Leica RM2155; Leica Instruments GmbH, Nussloch, Germany) and stained by haematoxylin and eosin method. Villi height and crypt depth were evaluated under a microscope using an image analysis programme (Image Pro Plus; Media Cybernetics, Bethesda, MD, USA). Caecal pH was measured directly using a Crison MicropH 2001 pH meter (Crison Instruments, Barcelona, Spain). The caecal content was immediately placed in sterile plastic tubes under ice for enzyme preservation and kept at −20 °C for further analysis of caecal enzyme activity.

2.5. Caecal Enzyme Activity

The crude enzyme extracted from the caecal content was extracted by homogenized caecal content with phosphate buffer solution (pH 7) at a 1:5 ratio (*w*/*v*). The homogenates were centrifuged at 18,000× *g* for 30 min at 4 °C to obtain the supernatant used to evaluate amylase, protease and cellulase activity. Amylase and cellulase activity were assayed according to [15,16] using 5% soluble starch and 1% carboxyl-methyl cellulose (CMC; medium viscosity) as the substrate, respectively. One hundred microlitres of crude enzyme extract were added to activate the digestion of the substrates. The products of the carbohydrate-digestive enzymes were stained using 1% dinitrosalicylic acid (DNS) and measured using a spectrophotometer at 540 nm against a linear range of maltose standards for amylase and glucose standards for cellulase. Protease activity was assayed according to the method described by [17] using 0.6% casein as the substrate. The product of the protein-digesting enzyme was measured spectrophotometrically at 660 nm against a linear range of tyrosine. The activity of the observed digestive enzymes was expressed as U.

2.6. Crude Protein Assessment

Each rabbit's feed intake between 19–24, 25–30 and 31–36 days old and the amount of crude protein in the milk replacer and diet were used as information to calculate the amount of crude protein intake.

2.7. Statistical Analysis

The results of this study are represented as the mean and pooled standard error of the mean. A completely randomized design was employed in this study. Therefore, one-way analysis of variance (ANOVA) was used to compare the different types of milk replacers (fixed factors) for internal organ characteristics, caecal pH, caecal digestive enzyme activities and duodenal histology, whereas the growth performances, water intake, faeces excretion, apparent digestibility, number of faecal pellets and amount of protein intake were analyzed by two-way mixed analysis of variance, with treatment groups or age serving as the between-subjects or the within-subjects factor, respectively. Duncan's multiple range test was used for post hoc analysis. Differences were considered statistically significant at $p < 0.05$. All statistical analyses in this study were performed with R-statistic software using the Rcmdr package [18].

3. Results

The effects on performance, apparent digestibility, amount of faeces excretion and crude protein intake from rabbits fed the different milk replacers are shown in Table 2. No statistically significant differences between the groups and the interactions between the studied factors (groups and age) for all parameters in Table 2 were apparent ($p > 0.05$). The age increment was correlated with increased body weight, ADFI, ADG, FCR, water intake, faeces excretion and crude protein intake ($p < 0.05$), whereas apparent digestibility did not affect dry matter, organic matter or ether extract ($p > 0.05$). However, the crude protein digestibility of rabbits at 31–35 days old was lower than rabbits at 23–27 days old ($p < 0.05$). The rabbits fed KMR and MMR displayed higher nutrient digestibility in both age ranges compared with rabbits fed RMR, but there was no statistically significant difference ($p > 0.05$). Rabbits in RMR, KMR and MMR were received the crude protein from milk daily at 2.23, 2.61 and 1.62 g dry matter/head. No deaths, morbidities or clinical signs were observed in rabbits during the experimental period. Moreover, a rabbit in each replicate was not euthanized at the end of the experiment and remained alive until it reached two months of age. In addition, there were no problems of milk perception and palatability in any group in this experiment, because the rabbits sucked milk directly from the syringe without any force feeding.

Table 2. Effect of different milk replacers on rabbit performances, apparent digestibility and crude protein intake.

Parameters	Factors							SEM	*p*-Value		
	Artificial Milk Replacers (AMR)			Age (days)					AMR	Age	AMR * Age
	RMR	KMR	MMR	0	6	12	18				
BW (g/head)	210	198	202	134 [a]	157 [b]	214 [c]	308 [d]	9.086	0.38	0.001	0.94
				19–24	25–30	31–36	-				
ADFI (g/head/day)	22.1	26.7	22.1	7.47 [a]	22.1 [b]	39.6 [c]	-	2.239	0.59	0.001	0.56
ADG (g/day)	9.62	9.33	8.52	4.40 [a]	10.8 [b]	14.6 [c]	-	0.808	0.84	0.001	0.77
FCR	2.39	2.62	2.74	1.91 [a]	2.21 [a]	2.75 [b]	-	0.115	0.44	0.01	0.56
Water intake (g/head/day)	41.8	30.9	34.2	10.1 [a]	31.9 [b]	68.0 [c]	-	3.902	0.1	0.001	0.38
Faeces excretion (g/head/day)	15	14.3	15.4	1.29 [a]	5.49 [b]	8.14 [c]	-	0.936	0.81	0.001	0.32
Crude protein intake (g/head/day) [1]											
Diet	3.72	3.73	3.9	1.20 [a]	3.56 [b]	6.37 [c]	-	0.361	0.59	0.001	0.56
Diet and milk	5.95	6.34	5.52	3.39 [a]	5.65 [b]	8.56 [c]	-	0.364	0.11	0.001	0.56
				23–27	31–35	-	-				
Apparent digestibility (%)											
Dry matter	59.6	63.3	63.5	60.5	63.1		-	1.126	0.29	0.22	0.96
Organic matter	60.7	65.1	65.2	62.8	64.2	-	-	1.123	0.18	0.51	0.85
Ether extract	68.3	73.8	70.7	71.3	70.3	-	-	2.418	0.75	0.88	0.97
Crude protein	76.5	80.8	81.1	81.5 [b]	77.1 [a]	-	-	1.007	0.07	0.03	0.34

RMR = Rabbit milk replacer, which was used in this study, KMR = Kitten milk replacer (KMR®, Pet-Ag Inc., Hampshire, IL, USA), MMR = Mammal milk replacer (Zoologic® Milk matrix 30/52, Pet-Ag Inc., Hampshire, IL, USA), SEM = pooled standard error of mean, BW = Body weight, ADFI = Average daily feed intake, ADG = Average daily weight gain, FCR = Feed conversion ratio; [a, b, c, d] The differences in superscript letter in the same row represented statistical significant differences ($p < 0.05$); [1] The rabbits in the RMR, KMR and MMR groups received crude protein from milk at 2.23, 2.61 and 1.62 g/head/day, respectively.

The consequences for internal organ weight, caecal pH, duodenal wall histology and digestive enzyme activities between the groups are compared in Table 3. The weight of each internal organ was not statistically significantly different between the groups ($p > 0.05$). Caecal pH was not influenced by the differences in milk replacers ($p > 0.05$). Caecal amylase activity in the MMR group was lower than in the KMR group ($p < 0.05$), whereas greater caecal protease activity was observed in the RMR group compared to the MMR group ($p < 0.05$). Cellulase activity was not affected by the treatments ($p > 0.05$). Respectively, the shortest villus and the shallowest crypt depth were found in the RMR and the KMR groups compared to the MMR group ($p < 0.05$).

Table 3. Effect of different milk replacers on internal organ weight, caecal pH, intestinal villi morphology and digestive enzyme activity.

Parameters	Artificial Milk Replacers (AMR)			SEM	*p*-Value
	RMR	KMR	MMR		
	Internal organs characteristics (g/live body weight)				
Liver	4.46	4.00	3.81	0.162	0.25
Spleen	0.14	0.10	0.10	0.015	0.51
Kidney	1.24	1.18	1.12	0.046	0.61
Thoracic organs [1]	1.07	1.09	1.15	0.063	0.38
Pancreas	0.06	0.07	0.05	0.007	0.99
Full stomach	7.40	7.44	7.22	0.555	0.79
Stomach wall	2.09	2.01	1.99	0.061	0.65
Intestinal organs [2]	7.38	6.87	7.37	0.242	0.54
Full caecum	16.3	16.1	16.2	0.422	0.97
Caecal wall	2.75	2.49	2.34	0.134	0.48
Caecal pH	6.50	6.30	6.38	0.081	0.63
	Caecal digestive enzyme activities (U)				
Amylase	12.2 ^{ab}	18.5 ^{b}	10.4 ^{a}	1.41	0.04
Protease (×10^{-1})	7.26 ^{b}	6.84 ^{ab}	5.79 ^{a}	0.253	0.04
Cellulase	3.71	3.50	3.60	0.057	0.31
	Duodenal villi morphology (µm)				
Villus height	320 ^{a}	361 ^{ab}	380 ^{b}	8.50	0.04
Villus crypt	66.2 ^{ab}	62.6 ^{a}	68.4 ^{b}	0.737	0.007

RMR = Rabbit milk replacer, which was used in this study, KMR = Kitten milk replacer (KMR®, Pet-Ag Inc., Hampshire, IL, USA), MMR = Mammal milk replacer (Zoologic® Milk matrix 30/52, Pet-Ag Inc., Hampshire, IL, USA), SEM = pooled standard error of mean, BW = Body weight, FI = Feed intake, ADG = Average daily weight gain, FCR = Feed conversion ratio; $^{a, b}$ The differences in superscript letter in the same row represented statistical significant differences ($p < 0.05$); [1] Thoracic organs includes lungs and heart; [2] Intestinal organs includes stomach, small intestine, large intestine, caecum and rectum with content.

The average number of faeces pellets from two rabbits of each replicate in the experiments is illustrated in Figure 1 and Table A1. The graph shows a steady increase in the number of faecal pellets with increasing age ($p < 0.001$); however, a sharp drop occurred in all study groups at 35 days of age, followed by another increase. The largest number of faecal pellets existed in the MMR group compared to the RMR group ($p < 0.05$; Appendix A Table A1), with the KMR group between them ($p > 0.05$). A significant interaction between the fixed factors (age and treatment group) was identified ($p < 0.05$). Generally, the same increasing trend was observed in all groups, except for the sharp rise in the number of faecal pellets in the MMR, RMR and KMR groups at 22–23, 24–26 and 34–36 days of age, respectively.

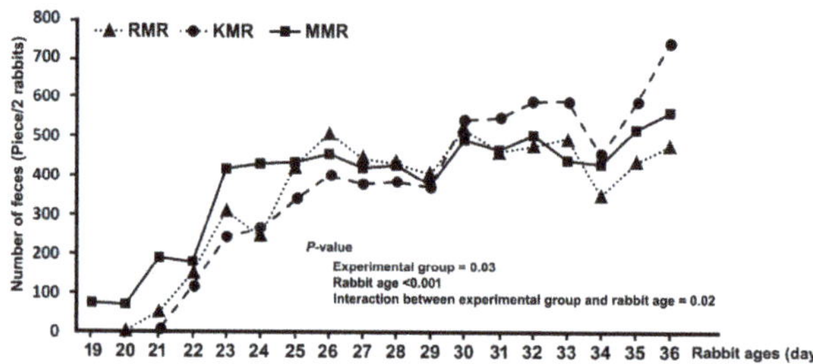

Figure 1. Effect of different milk replacers on number of faecal pellets (RMR, rabbit milk replacer in this study; KMR, kitten milk replacer, KMR®, Pet-Ag Inc., Hampshire, IL, USA; and MMR, mammal milk replacer, Zoologic® Milk matrix 30/52, Pet-Ag Inc., Hampshire, IL, USA).

4. Discussion

The RMR was formulated according to the profile of rabbit milk; therefore, its chemical composition was the most similar to the composition of rabbit milk [10]. KMR contained a higher proportion of crude protein and was lower in fat and energy than RMR and rabbit milk. On the other hand, MMR was lower in crude protein and higher in fat and energy than RMR and rabbit milk. A high density of nutrients and energy was the unique characteristic of rabbit milk, which contained respectively around four and three times higher proportions of protein and lipids than cow's milk [10]. A short milking time is a common nursing behaviour, explaining the high density of rabbit milk [6]. The amount of milk replacer fed to young rabbits was calculated on the basis of stomach capacity. Milk replacer in powder form was used for all formulations in this study because a highly concentrated milk mixture can be formulated from dried powder but not in liquid form [8,9]. Rabbit milk protein comprises around 70% and 30% casein and whey protein, respectively [10]. Therefore, casein served as a major protein ingredient in the milk replacer formulation for RMR and the two other commercial milk replacers, whereas dried skimmed milk powder represented another protein source, which was used in a lower proportion than casein. Respectively, either whey or milk protein concentrate was supplemented in KMR and MMR, whereas in RMR they were not. A low level of lactose is present in rabbit milk; therefore, cow's or goat's milk is limited as a milk replacer formulation [10]. Moreover, lactase activity in rabbits decreases with age and does not respond to the lactose concentration in the diet. Therefore, excessive lactose intake can lead to a digestive disorder [6]. Differences in the type of raw protein source and quantity influence the diversity of the amino acid profile. Although the amino acid profile was not evaluated in this study, the intake of amino acids would have been sufficient for rabbits because casein, considered an ideal protein, was used as the main ingredient and the percentage of crude protein in all milk replacers was higher than nutrient requirements [4,6].

Fat in rabbit milk represents the major energy sources for rabbits [10], whereas excessive starch intake promotes digestive problems and increases the mortality rate of young rabbits [4]. The highest nitrogen-free extract was present in the KMR formulation; however, no adverse effects were observed in this group. Medium-chain fatty acids, mainly caprylic (C8:0) and capric acid (C10:0), were the major components of fatty acids in rabbit milk, comprising around 50% of total fatty acids [10]. Vegetable oil or hydrogenated palm oil served as the only lipid sources in milk replacers, comprising a high proportion of polyunsaturated fatty acids; however, they still contained these medium-chain fatty acids. Another function of caprylic and capric acid was their antibacterial properties, which maintained microbial community development and prevented pathogen invasion. Supplementation with these medium-chain fatty acids should confer health benefits on rabbits. Nevertheless, no health problems occurred in this study, although these medium-chain fatty acids were not supplied in the formulations. The essential fatty acids were supplied in sufficient amounts to fulfil the nutritional requirements of all the study groups, as there was a very high amount of fat in all the formulations [6]. However, an antioxidant such as tocopherol should be supplemented to prevent lipid oxidation. The minerals and vitamins in RMR were higher than the minimum requirements in rabbits [6]. In the early period of the experiment, the rabbits received nutrients and energy from the milk replacers in high proportions compared to the pelleted diet, which represented only around 1/3 of the total intake in the first period. Subsequently, a higher intake of the pelleted diet became the main nutrient and energy source. Growth performances did not differ between treatment groups, and no morbidity or mortality of rabbits occurred during or after the experimental period. Moreover, the rabbits' final body weights were similar to those of their counterparts reared with their mother at the same age, as reported in the study of [5]. These findings demonstrate the potential of using milk replacers in early weaned rabbits. However, the low energy and nutrient requirements of the rabbits in this study were characteristics of a native breed with a slow growth rate compared to hybrid or commercial meat rabbits [5]. Therefore, milk replacers can certainly be used in local breed or pet rabbits, possibly offering benefits to veterinarian and pet rabbit producers. Rabbits raised in intensive rearing systems and/or with high growth rates may be studied further as to which milk replacer can support their growth performances and the maintenance of a healthy condition. Another study [9] reported lower growth performances in commercial meat rabbits fed KMR compared with rabbits fed rabbit milk.

A previous report has shown that the gut microbial community of rabbits plays an important role relating to the efficiency of nutrient fermentation, productive performance and heath condition [19]. A sterile gut is observed in rabbits immediately after birth. Subsequently, microbes slowly and continuously colonize the rabbit gut and the amount of antimicrobial substances in rabbit milk and nutrients have major effects on the development of gut functions and the microbial community [10,19]. A large variation in the bacterial community was observed between young rabbits at an early age. However, the composition of the bacterial community between rabbits became more similar with increasing age, especially with pelleted diets [13,19]. To our knowledge, the caecum is a major organ of bacterial fermentation in rabbits, providing around 50–60% of daily energy requirements. The caecal environment greatly affects the microbial community in rabbits, especially in terms of pH, being influenced in a major way by fermentable products called volatile fatty acids. A low nutrient fermentation efficiency can result in a higher value of caecal pH, increasing the risk of a digestive disorder [6]. The caecal pH of the RMR group seemed to be higher than that of the other groups. However, the caecal pH of the RMR group was lower than 6.73, insufficient to promote a gut health problem [6]. Moreover, there was no statistically significant difference in caecal pH, growth performance, digestibility, morbidity and mortality during the experiments, supporting the possibility of using RMR in rabbits.

Caecal enzyme activity was another indicator representing microbial activity. The amount of caecal amylase and protease in the RMR and KMR groups was higher than in the MMR group. Fermentation products from probiotic bacteria and prebiotics play an important role in supporting the development of a normal flora in rabbits [20]. Supplementation with fermentation products from several normal flora bacteria (*Lactobacillus* sp., *Enterococcus faecium*, *Bifidobacterium bifidum*, *Pediococcus adicilactici*) and prebiotics (fructooligosaccharide) in the KMR formulation contributed to higher enzyme activity, whereas RMR and MMR did not contain these supplements. Such findings are in accordance with another research study [8]. A higher survival rate was observed in the milk replacer with probiotics and prebiotics (35.3%) compared to without these additions (21.3%) in infant cottontail rabbits in which the chemical composition of these milk replacers were similar [8]. Thus, probiotics and/or prebiotics may have been the key factor determining this result. The unsuitable chemical components of KMR with respect to rabbit milk and the slow growth rate of bacteria in the first period could have been the cause of the lower feed intake in the first period in the KMR group. Subsequently, feed intake in the KMR group was higher than in the others at the end of the experiment, as a consequence of the full development of microbes. The closeness of the chemical component of RMR to that of rabbit milk may have promoted the appropriate substrate for early microbial colonization, leading to higher caecal enzyme activity. None of prebiotic supplement and difference in chemical composition in MMR could be the cause of the lower enzyme activity in the caecum compared to other groups. Therefore, supplementation with probiotics and/or substrates for a normal flora in feed formulation with the appropriate chemical components (i.e., in line with rabbit milk) may represent the best procedure to achieve good development of the microbial community in early weaned rabbits. However, a microbial community analysis should be performed in a future study to confirm this hypothesis.

No statistically significant difference in growth performance was observed in this study. However, a trend existed that can be explained in detail. The lowest FCR was observed in the RMR group as a consequence of the lower ADFI and the higher ADG. This may have been due to the appropriate crude protein and fat proportion in the RMR, which promoted lower adaptation following the milk replacer in the first period. However, the highest ADFI was observed in the KMR group as the lowest energy density because the feed intake was stimulated by a physiological function until the animal had obtained its daily energy requirement [6]. Therefore, a lower pelleted diet intake could be observed in the group already receiving a high-density milk replacer in the RMR and MMR groups. The digestibility of dry matter and nutrients in the RMR group tended to be lower than in the other groups. The supplementation of prebiotics and substrates for microflora in KMR can support the microbial community and facilitate increased digestibility. In addition, MMR contained the highest energy density, which bacteria can use as energy sources to produce butyrate. Intestinal epithelial cells can utilize such short-chain fatty acids as an energy source, promoting proliferation, differentiation and gut immunity [21]. Thereby, the greatest villus height and villus crypt depth were seen in the MMR group, enabling better digestion and absorption and providing higher nutrient digestibility RMR as a result [22]. Furthermore, the amount and characteristics of hard faeces can be used to indicate digestive health [7]. The total weight of faecal excretion did not differ between groups, but the largest number of faecal pellets was found in MMR, followed by KMR and RMR. Moreover, no soft faeces were found under the rabbit cage, indicating that crude protein intake did not exceed their requirement [19].

Lower growth performance and survival rates were reported in the group fed with KMR compared with the group fed with rabbit milk from lactating does [8]. Unfortunately, this study did not compare the milk replacers with rabbit milk. Early separation at 14 days of age may have been the cause of the negative consequences found in the study of [8], whereas separation at 18 days of age did not result in any serious adverse outcomes here. In addition, a high mortality rate was observed in the study of [8] because the rabbits were too young, were injured and were experiencing high levels of stress due to being wild rabbits.

5. Conclusions

Differences in efficiency between the RMR developed in this study and commercial milk replacers (KMR and MMR) used in 18-day-old rabbits were revealed in the current study. Based on the results, it was possible to use RMR as a milk replacer for 18-day-old rabbits and to wean at 36 days of age, this not providing any adverse consequences for final body weight, ADG, FCR, ADFI, nutrient digestibility, internal organ characteristics, caecal pH, amount of faeces excretion and crude protein intake. Lower nutrient digestibility was observed in the RMR group without statistically significant differences. Therefore, probiotics and/or prebiotics can be supplemented in formulations to promote a suitable microbial community and to provide benefits in terms of growth performance and nutrient digestibility.

Author Contributions: Conceptualization, A.K.; methodology, A.K.; investigation, P.C. (Panthiphaporn Chankuang), A.L., K.J., C.K., T.S., P.C. (Pipatpong Chundang) and A.K.; resources, P.S. and C.Y.; data curation, A.K.; writing—original draft preparation, A.K.; writing—review and editing, P.C. (Panthiphaporn Chankuang), A.L., K.J., C.K., T.S., P.C. (Pipatpong Chundang), A.K., P.S. and C.Y.; project administration, A.K.; funding acquisition, A.K. All authors have read and agreed to the published version of the manuscript.

Funding: This research was funded by Student Development Fund, Faculty of Veterinary Medicine, Kasetsart University, Bangkok, Thailand.

Acknowledgments: The authors would like to thank Faculty of Veterinary Medicine, Veterinary Teaching Hospital Kamphaengsaen campus (Kasetsart University, Nakhon Pathom, Thailand) and Mr.Saha Bairak (Saha farm, Kanchanaburi, Thailand) for providing the rabbit milk replacer and rabbits, respectively. Moreover, we would like to acknowledge Thanaporn Sriprathardtrakul, Jiraphat Wangka and Sirapoom Narktap for the support in rabbit rearing and feeding.

Conflicts of Interest: The authors declare no conflict of interest.

Appendix A

Table A1. Effect of different milk replacers on number of faecal pellets.

Artificial Milk Replacers	Age (days)	Number of Feces
RMR [A]		
	19	0 [a]
	20	0 [a]
	21	6 [ab]
	22	117 [b]
	23	244 [cd]
	24	263 [c]

Table A1. *Cont.*

Artificial Milk Replacers	Age (days)	Number of Feces
	25	338 ^{def}
	26	401 ^{efgh}
	27	379 ^{defg}
	28	385 ^{defg}
	29	371 ^{cde}
	30	543 ^{hi}
	31	548 ^{fgh}
	32	590 ^{hi}
	33	592 ^{ghi}
	34	454 ^{def}
	35	587 ^{ghi}
	36	742 ^{i}
KMR ^{AB}		
	19	75.6
	20	70.8
	21	191
	22	180
	23	418
	24	432
	25	435
	26	456
	27	421
	28	428
	29	378
	30	493
	31	467
	32	505
	33	441
	34	431
	35	519
	36	563
MMR ^{B}		
	19	0
	20	3.33
	21	52.3
	22	151
	23	311
	24	245
	25	426
	26	509
	27	444
	28	432
	29	406
	30	520
	31	463
	32	476
	33	497
	34	350
	35	436
	36	477
SEM		12.02
p-Value		
	Groups	0.03
	Age	0.001
	Groups * Age	0.02

RMR = Rabbit milk replacer in this study, KMR = Kitten milk replacer (KMR®; Pet-Ag Inc., Hampshire, IL, USA), MMR = Mammal milk replacer (Zoologic® Milk matrix 30/52; Pet-Ag Inc., Hampshire, IL, USA; SEM = Pooled standard error of mean; $^{a–i, A, B}$ The differences in superscript capital or lower case letter in the same column represent statistical significant differences of groups or age, respectively ($p < 0.05$).

References

1. Grand View Research. Pet Care Market Size, Share & Trends Analysis Report by Pet Type (Dog, Cat, Fish, Bird), by Products, Competitive Landscape, and Segment Forecasts, 2018–2025. Available online: https://www.grandviewresearch.com/industry-analysis/pet-care-market (accessed on 20 April 2020).
2. Morley, C.; Fook, J. The importance of pet loss and some implications for services. *Mortality* **2005**, *10*, 127–143. [CrossRef]
3. Rooney, N.J.; Blackwell, E.J.; Mullan, S.M.; Saunders, R.; Baker, P.E.; Hill, J.M.; Sealey, C.E.; Turner, M.J.; DE Held, S. The current state of welfare, housing and husbandry of the English pet rabbit population. *BMC Res. Notes* **2014**, *7*, 942–954. [CrossRef] [PubMed]
4. Gutiérrez, I.; Espinosa, A.; Gracía, J.; Carabano, R.; De Blas, J.C. Effect of levels of starch, fiber, and lactose on digestion and growth performance of early-weaned rabbits. *J. Anim. Sci.* **2002**, *80*, 1029–1037. [CrossRef] [PubMed]
5. Kovitvadhi, A.; Chandang, P.; Tirawattanawanich, C.; Kasemsuwan, S.; Rukkwamsuk, T. The study on cluster and obstacles of meat rabbit production in Thailand. *J. Mahanakorn. Vet. Med.* **2019**, *14*, 115–127.
6. De Blas, C.; Wiseman, J. *Nutrition of the Rabbit*, 2nd ed.; CAB International: Oxfordshire, UK, 2010; pp. 1–19, 180–185, 229.
7. Varga, M. *Textbook of Rabbit Medicine*, 2nd ed.; Butterworth Heinemann: London, UK, 2014; pp. 84–87.
8. Paul, G.; Friend, D.G. Comparison of outcomes using two milk replacer formulars based on commercially available products in two species of infant cottontail rabbits. *J. Wildl. Rehabil.* **2017**, *37*, 13–19.
9. Ferguson, F.A.; Lukefahr, S.D.; McNitt, J.L. A technical note on artificial milk feeding of rabbit kits weaned at 14 days. *World Rabbit Sci.* **1997**, *5*, 65–70. [CrossRef]
10. Maertens, L.; Lebas, F.; Szendrö, Z.S. Rabbit milk: A review of quantity, quality and non-dietary affecting factors. *World Rabbit Sci.* **2006**, *14*, 205–230. [CrossRef]
11. AVMA (American Veterinary Medical Association). *AVMA Guidelines for the Euthanasia of Animals: 2020 Edition*; American Veterinary Medical Association: Schaumburg, IL, USA, 2020; pp. 63–64.
12. AOAC (Association of Official Analytical Chemists). *Official Method of Analysis*, 18th ed.; Association of Official Analytical Chemists: Rockville, MD, USA, 2006.
13. AFFCO (Association of American Feed Control Officials). *2020 Official Publication*; Association of American Feed Control Officials Inc.: Champaign, IL, USA, 2020.
14. Kovitvadhi, A.; Gasco, L.; Ferrocino, I.; Rotolo, L.; Dabbou, S.; Malfatto, V.; Gai, F.; Peiretti, P.G.; Falzone, M.; Vignolini, C.; et al. Effect of purple loosestrife (*Lythrum salicaria*) diet supplementation in rabbit nutrition on performance, digestibility, health and meat quality. *Animal* **2016**, *10*, 10–18. [CrossRef]
15. Areekijseree, M.; Engkagul, A.; Kovitvadhi, U.; Thongpan, A.; Mingmuang, M.; Pakkong, P.; Rungruangsak-Torrissen, K. Temperature and pH characteristics of amylase and proteinase of adult fresh water pearl mussel, *Hyriopsis* (*Hyriopsis*) *bialatus*, Simpson (1900). *Aquaculture* **2004**, *234*, 575–578. [CrossRef]
16. Vatanparast, M.; Hosseininaveh, V.; Ghadamyari, M.; Minoo-Sajjadian, S. Plant cell wall degrading enzymes, pectinase and cellulase, in the digestive system of the red palm weevil, *Rhynchophorus ferrugineus* (Coleoptera: Curculionidae). *Plant Prot. Sci.* **2014**, *50*, 190–198. [CrossRef]
17. Folin, O.; Ciocalteu, V. On tyrosine and tryptophan determinations in proteins. *J. Biol. Chem.* **1927**, *73*, 627–650.
18. R: A Language and Environment for Statistical Computing. Available online: https://www.R-project.org (accessed on 19 March 2020).
19. Combes, S.; Fortun-Lamothe, L.; Cauquil, L.; Gidenne, T. Engineering the rabbit digestive ecosystem to improve digestive health and efficacy. *Animal* **2013**, *7*, 1429–1439. [CrossRef] [PubMed]
20. Falcão-e-Cunha, L.; Castro-Solla, L.; Maertens, L.; Marounek, M.; Pinheiro, V.; Freire, J.; Mourão, J.L. Alternatives to antibiotic growth promoters in rabbit feeding: A review. *World Rabbit Sci.* **2007**, *15*, 127–140. [CrossRef]
21. Kinoshita, M.; Suzuki, Y.; Saito, Y. Butyrate reduces colonic paracellular permeability by enhancing PPARγ activation. *Biochem. Biophys. Res. Commun.* **2002**, *293*, 827–831. [CrossRef]

22. Makovicky, P.; Tumova, E.; Volek, Z.; Makovicky, P.; Vodicka, P. Histological aspects of the small intestine under variable feed restriction: The effects of short and intense restriction on a growing rabbit model. *Exp. Med.* **2014**, *8*, 1623–1627. [CrossRef]

Review

Autochtonous Strain *Enterococcus faecium* EF2019(CCM7420), Its Bacteriocin and Their Beneficial Effects in Broiler Rabbits—A Review

Monika Pogány Simonová [1,*], Ľubica Chrastinová [2] and Andrea Lauková [1,*]

1 Institute of Animal Physiology, Centre of Biosciences of Slovak Academy of Sciences, Šoltésovej 4-6, 04001 Kosice, Slovakia

2 Institute for Nutrition, National Agricultural and Food Centre, Hlohovecká 2, 951 41 Nitra-Lužianky, Slovakia; lubica.chrastinova@nppc.sk

* Correspondence: simonova@saske.sk (M.P.S.); laukova@saske.sk (A.L.); Tel.: +421-55-792-2964 (M.P.S.); +421-55-792-2964 (A.L.)

Received: 22 June 2020; Accepted: 10 July 2020; Published: 14 July 2020

Simple Summary: Weaning is the most important and critical period in rabbits breeding; the cecal digestion is very complex and only small dietary and/or environmental changes can disturb the stable microbial population/fermentation and gut health, leading to digestive dysbiosis and increased morbidity, often with fatal outcome and big economic losses. Control of the microbiota, prevention of digestive disturbances and improving gut health and immunity can be achieved through the natural substances application in rabbit nutrition. While probiotics are frequently used in rabbit farms, the in vivo administration of bacteriocins (antimicrobial substances produced by bacteria, which usually also possess probiotic properties) in these animals is often limited and has become an area of research activity. Moreover, the most of probiotic strains used in rabbits are non-autochthonous (have a different origin than the rabbits ecosystem). Therefore, our study focused on improving rabbits' health using the autochthonous strain *Enterococcus faecium* EF2019 (CCM7420) and its enterocin (Ent7420) in broiler rabbits. The antibacterial and anticoccidial effect of additives was observed, with good colonization ability of the CCM7420 strain. Both additives showed a tendency to modulate the serum biochemistry parameters and to improve the immunity, jejunal morphology, weight gains, feed conversion ratio and meat quality (physicochemical traits and mineral content).

Abstract: The present review evaluates and compares the effects achieved after application of rabbit-derived bacteriocin-producing strain *Enterococcus faecium* CCM7420 with probiotic properties and its bacteriocin Ent7420. The experiments included varying duration of application (14 and 21 days), form of application (fresh culture and lyophilized form), combination with herbal extract and application of the partially purified enterocin—Ent7420, produced by this strain. Results from these studies showed that *E. faecium* CCM7420 strain was able to colonize the gastrointestinal tract (caecum) of rabbits (in the range < 1.0–6.7 log cycle, respectively 3.66 log cycle on average), to change the composition of intestinal microbiota (increased lactic acid bacteria, reduced counts of coliforms, clostridia and staphylococci), to modulate the immunity (significant increase of phagocytic activity), morphometry (enlargement absorption surface in jejunum, higher villi height:crypt depth (VH:CD) ratio), physiological (serum biochemistry; altered total proteins, glucose and triglycerides levels) and parasitological (*Eimeria* sp. oocysts) parameters and to improve weight gains (in the range 4.8–22.0%, respectively 11.2% on average), feed conversion ratio and meat quality (physicochemical traits and mineral content).

Keywords: rabbit; *Enterococcus faecium*; enterocin; microbiota; intestinal morphology; phagocytic activity; serum biochemistry; meat quality; weight gain

1. Introduction

Rabbit breeding has a great potential because of the small body size, short generation interval, rapid growth rate, high productive capacity and healthy, easily digestible meat of rabbits [1,2]. Moreover, rabbits can convert a higher amount (20%) of the protein they eat into edible meat, compared with pigs (16–18%) and cattle (8–12%; [3]). In several European countries, in which the rabbit breeding has a long history and high production efficiency, presently is regressing, whereas in the developing countries of the world rabbit farming has become to an important emerging enterprise. In growing rabbits, the most critical period is the weaning, when the kits are separated from mother and the milk is substituted with solid feed [4]. During these environmental and physiological changes/stresses, the rabbits are very sensitive to digestive disturbances, also called non specific enteritis (usually caused by dietary stresses, parasites—*Coccidia* and bacteria—*Clostridia* sp. and enteropathogenic *Escherichia coli*) and gastrointestinal infections (epizootic rabbit enteropathy—ERE, a multifactorial gastrointestinal syndrome, [5]). These dietary and bacterial changes are the main reason of morbidity and mortality and have negative effects on feed consumption, growth performance and health status of animals in this period and also on the economic aspects of rabbit farming.

To overcome this period, to reduce economic losses and to improve and stabilize the health status and gastrointestinal tract development, antibiotic growth promoters (AGPs) have been widely used for years. Although, these synthetic drugs showed good effects on production indicators, on the other hand, there was a risk of increasing resistance to antibiotics and transferring of antibiotic resistance genes from animal to human, which also threatened the human health and quality of meat and food [6]. For this reason, AGPs have been banned by the European Union (began in 1986 in Sweden and completely banned in January 2006, when the last four antibiotics have been permitted as feed additives was no longer allowed to be marketed or used from this date; IP/03/1058; [7,8]). As a result of the ban, researchers had to substitute AGPs and to find new feed additives that were supposed to be safer, without leaving residues and spreading resistance to themselves, but also improving health and productivity of rabbits. Therefore, the antibiotics have been replaced with new, naturally based supplements: probiotics, prebiotics, synbiotics, enzymes, bacteriocins, organic acids, herbs and their extracts, which are well-tried tools for disease prevention and therapy in various animal species, including rabbits [9].

The last two decades have seen a substantial increase in the use natural supplements and/or additives in animal nutrition, in which their antimicrobial activity has been highlighted many times. The EU in its Regulation EC 1831/2003 defined the terms "feed additives" as "substances, microorganisms or preparations, other than feed material and premixtures, which are intentionally added to feed or water" and the "antimicrobials" as "substances produced either synthetically or naturally, used to kill or inhibit the growth of microorganisms, including bacteria, viruses or fungi, or of parasites, in particular protozoa" [10]. According to the World Health Organization (WHO) probiotics are defined as "live microorganisms which, when administered in adequate amount, confer a health benefit on the host". Although this definition is widely accepted, a 2007 guidance document from the European Commission on Regulation EC 1924/2006 on nutrition and health claims (NHCR) categorizes the term 'probiotic' as a health claim on the basis that it implies a health benefit. For this reason, the term "beneficial microbes" is used more often instead of a "probiotic microorganism" [11].

The use of several natural feed additives has already been reviewed in rabbit breeding [10,12–15]. The positive effects of probiotics and their antibacterial products—bacteriocins on health, growth performance, nutrient utilization and metabolism changes, microbial composition [14,16–35], blood serum biochemistry, oxidative stress, immune response, intestinal morphology [21,24,28–31,34,36–38] and meat quality of rabbits [32,34,39–43] was described. However, in spite of the achieved results, there are still few declared probiotic preparations based (Lactina, Toyocerin; [44,45]; Prorabbit—declared in Slovakia; [25]) and detailed studies of microorganisms with beneficial properties that have also the ability to produce antimicrobial substances, enterocins.

The aim of this review was to summarize all achieved properties and physiological effects of the bacteriocin-producing strain with probiotic properties *Enterococcus faecium* CCM7420 (EF2019 previous working labeling, [46]) isolated in 2003 from rabbit feces in the Laboratory of Animal Microbiology of the Institute of Animal Physiology, Centre of Biosciences of the Slovak Academy of Sciences (Košice, Slovakia) and tested to date in 180 rabbits. These experiments included varying duration of application (2 and 3 weeks), form of application (fresh culture in water; the concentration of cells was $\times 10^9$ CFU/mL in a dose 500 μL/animal/day; lyophilized (freeze-dried) form rehydrated in water ($\times 10^9$ CFU/mL; dose 500 μL/animal/day) as well as mixed in feed and pelleted (15 g/100 kg feed), application of its partially purified bacteriocin (PPB)—enterocin (Ent) EF2019 (applied into water) and fresh culture in combination with natural substance (*Eleutherococcus senticosus*).

2. *Enterococcus faecium* CCM7420 (EF2019) and Its Bacteriocin-Enterocin (Ent7420)

Enterococcus faecium EF2019 (CCM7420) is a bacteriocin-producing strain [47], which was isolated from the rabbit feces and genetically confirmed by the PCR method and subsequently by MALDI-TOF mass spectrophotometry as well as the sequencing procedure of this strain was provided (Dr. Kopčáková, IAP CBs SAS). This strain produces lactic acid, tolerates low pH (3.0; 63% surviving of cells) and is able to grow even in 5% oxgall—bile (80% surviving of cells), shows sensitivity to antibiotics, including vancomycin [25,48] and possess lipolytic activity [49]. Other unpublished data suggests that the CCM7420 does not produce biogenic amines and enzymes such as β-glucuronidase, β-galactosidase or N-acetyl-β-glucosaminidase (enzymes produced by unfriendly gut bacteria; their increased levels are usually the indicators of colon cancer), and it does not show any gelatinase (absence of the *gelE* gene) or hemolytic activities with low ability to form biofilm (0.092). The strain was deponed into Czech Collection of Microorganisms in Brno, Czech Republic to have number CCM7420. This strain showed the broadest inhibitory activity from all tested rabbits enterococcal strains against the indicators *E. avium* EA5, *Listeria innocua* LMG13568 and *L. monocytogenes* CCM4699 and against other tested enterococci and staphylococci tested such as clostridia, pseudomonads, enterobacteria and coliform bacteria [48]. The presence of the structural genes for enterocins (ent) A, P and L50B was detected; however, the CCM7420 did not possessed gene for ent B [47]. The molecular mass of its bacteriocin-like substance ranged from 3 to 10 kDa. Proteinaceous substance produced by CCM7420 strain was partially purified (partially purified bacteriocin (PPB) or enterocin (Ent) 2019 =7420). It is thermostable substance as well as stable at pH 4.0, 7.0 and 9.0. Its production starts in early logarithmic growth phase and it culminates in the late logarithmic phase of CCM7420 strain growth. By its properties, it can probably be included in the II. classification group of bacteriocins. Ent2019 or Ent7420 added to the growing strain *L. innocua* LMG13568 (after 4 h) inhibited its growth already at 1 h after enterocin addition with a difference of 1.5 log cycles (5 h of cultivation). This effect was prolonged up to 24 h. The Ent7420 was tested against more than 300 strains of enterococci, staphylococci, clostridia, pseudomonads, enterobacteria and coliforms [48]. The inhibitory activity of this substance was preserved after 24 months of storage at −20 °C (6400 AU/mL; [48]) and also after lyophilization (freeze dried) and redissolution in PBS buffer (25600 AU/mL; not published data). The CCM7420 strain is currently available in the ProRabbit, probiotic product for rabbits and other rodents, made by the International Probiotic Company Košice (Slovakia).

3. Application Effects of *E. faecium* CCM7420 and Its Enterocin Ent7420 Observed in Experiments

3.1. *Effect on Growth Performance*

The results from all these experiments indicated that the CCM7420 strain could improve the average daily weight gain (ADWG; between 4.8 and 22.0%; Table 1) regardless of the form of bacteriocin-producing strain (fresh, $p < 0.001$ compared to control data; or freeze dried-lyophilized—numerical increase) and its application time (2 or 3 weeks). The feed conversion ratio (FCR) was influenced only through the Ent7420 administration and the combinative application of the CCM7420 strain with *Eleutherococcus senticosus* extract. The effect of probiotics, including registered probiotic preparations and new beneficial microorganisms on the growth performance of rabbits have been already described/reviewed; they usually confirmed the increased body weight [14,15,17,19,24,25,31,32]. The most of these studies described faster growth and higher weight gain of rabbits using probiotic preparations and feed additives based on the following bacterial strains and yeasts alone or in their combinations: *Saccharomyces cerevisiae, S. boulardii, Bacillus licheniformis, B. cereus, B. cereus* var. *toyoi, Pediococcus acidilactici, Lactococcus lactis, Lactobacillus acidophilus, L. plantarum, L. helveticus, L. delbrueckii* and *L. sporogenes* [14]. Up to now, only several commercial products recommended especially for rabbits contain especially the species strain *Enterococcus faecium* as a component of the probiotic bacterial mix (Lactina—*E. faecium* NBIMCC 8270 [44]; Prorabbit—*E. faecium* CCM7420 [25]) or for companion animals with diarrhea (Pro-enteric Triplex—*E. faecium* DSM 10663/NCIMB 10415 [50]). In all experiments, higher ADWG was noted during CCM7420 administration. Surprisingly, the highest increase of ADWG (by 22%, Table 1) was noted after 2 weeks addition (model experiment, only seven animals were used in the group). The three weeks long CCM7420 dietary inclusion also improved the growth performance; however, only by 6.7% on average (from results of fresh culture application in two experiments; [25,26]). Similarly to our results, Lauková et al. [21] and Szabóová et al. [35] also reported higher ADWG through bacteriocinogenic and probiotic *E. faecium* AL41 (CCM8558) and CCM4231 strains application in rabbits. Improved body weight in rabbits was also noted after probiotics administration by Bovera et al. [30], Bhatt et al. [32] and Kalma et al. [51]. On the other hand, after PPB CCM7420 addition increased the body weight only slightly (by 2.2%, Table 2), but in this group, better FCR was achieved, compared to strain application. The bacteriocin dietary inclusion showed better feed conversion also in the case of other bacteriocins: EntM, EntCCM4231, EntEF55 and gallidermin [21,52–54], applied in rabbit husbandry.

Table 1. The effect of *Enterococcus faecium* CCM7420 and its enterocin Ent7420 on the growth performance of rabbits.

Tested Zootechnical Parameters	*E. faecium* CCM7420 Strain					Enterocin (Ent) 7420
	Fresh Culture	Fresh Culture	Lyophilized Form Resolved in Water	Lyophilized Form Mixed into Pellets	Fresh Culture + *E. senticosus*	
Reference number of publication	[24]	[25]	[55]	[55]	[26]	[25]
Length of application	14 days	21 days	21 days	21 days	21 days	21 days
Number of rabbits	(n = 7)	(n = 24)	(n = 24)	(n = 24)	(n = 24)	(n = 24)
Initial live weight (35 d of age; 0 d of experiment), g	1136.0 ± 100.0	977.0 ± 97.0	1002.3 ± 162.3	1042.5 ± 315.7	1077.5 ± 102.2	963.0 ± 101.0
Intermediate live weight (49/56 d of age; 14/21 d of experiment), g	1507.0 ± 120.0	1850.0 ± 152.0	1664.4 ± 170.0	1856.9 ± 361.4	1943.3 ± 222.2	1788.0 ± 199.0
Final weight (70/77 d of age; 35/42 d of experiment), g	2325.0 ± 260.0	2622.0 ± 104.0	2206.7 ± 164.6	2319.2 ± 164.6	2723.3 ± 204.7	2431.0 ± 142.0
Average daily weight gain (g/d; increase compare to control data%)	28.00 (22.0%)	39.17 (4.8%)	38.35 (10.0%)	38.51 (10.5%)	39.19 (8.5%)	34.95 (2.2%)
Feed conversion ratio between 35 and 56 days of age (g/g)	Not tested	2.71	2.23	2.28	2.59	2.68
Feed conversion ratio between 56 and 77 days of age (g/g)	Not tested	4.67	3.41	3.42	3.97	3.87
Feed conversion ratio per kg gain	Not tested	3.47	2.82	2.85	3.28	3.22
Mortality (n(%) in experimental group/n(%) in control group)	0(0.0%)/0(0.0%)	3(12.5%)/5(20.8%)	3(12.5%)/7(29.2%)	3(12.5%)/7(29.2%)	0(0.0%)/4(16.7%)	1(0.04%)/4(16.7%)

The fresh culture of the CCM7420 strain was applied into water (at concentration of cells $\times 10^9$ CFU/mL; dose 500 μL/animal/day); lyophilized (freeze-dried) form rehydrated in water ($\times 10^9$ CFU/mL; dose 500 μL/animal/day) as well as mixed in feed and pelleted (15 g/100 kg feed).

Table 2. The effect of *Enterococcus faecium* CCM7420 and its enterocin Ent7420 on the fecal microbiota of rabbits.

	E. faecium CCM7420 Strain					Enterocin (Ent)7420
Tested Microorganisms	**Fresh Culture**	**Fresh Culture**	**Lyophilized form Resolved in Water**	**Lyophilized Form Mixed into Pellets**	**Fresh Culture + *E. senticosus***	
Reference number of publication	[24]	[25]	[55]	[55]	[26]	[25]
Length of application	14 days	21 days	21 days	21 days	21 days	21 days
Enterococci	Increased	Increased ($p < 0.01$)	Increased	Increased	Unchanged	Increased ($p < 0.05$)
Lactic acid bacteria (LAB)	Increased ($p < 0.01$)	Unchanged	Increased	Increased	Unchanged	Decreased
Clostridia	Decreased	Decreased	Unchanged	Unchanged	Decreased ($p < 0.05$)	Unchanged
Coagulase-positive staphylococci	Decreased ($p < 0.01$)	Unchanged	Unchanged	Unchanged	Decreased	Decreased
Staphylococcus aureus	Unchanged	Decreased	Unchanged	Unchanged	Decreased ($p < 0.001$)	Decreased
Coliforms	Unchanged	Decreased	Unchanged	Unchanged	Decreased ($p < 0.001$)	Decreased
CCM7420	6.7 log cycle	6.7 log cycle	<1.0 log cycle	2.8 log cycle	1.1 log cycle	-

The fresh culture of the CCM7420 strain was applied into water (at concentration of cells $\times 10^9$ CFU/mL; dose 500 µL/animal /day); lyophilized (freeze-dried) form rehydrated in water ($\times 10^9$ CFU/mL; dose 500 µL/animal/day) as well as mixed in feed and pelleted (15 g/100 kg feed). Statistical analysis was performed using one-way analysis of variance (ANOVA) with the post hoc Tukey test with the level of significance set at ($p < 0.05$), within experimental groups during each individual experiment.

3.2. *Effect on Fecal Microbiota*

In all experiments, administration of *E. faecium* CCM7420 was associated with increased enterococci ($p < 0.01$) and lactic acid bacteria (LAB) counts during the treatment period by a 1.3 log cycle on average (Table 2), while the application form or time length had no impact on the bacterial counts increase. The numerical increase of enterococci and LAB was recorded also in previous experiments with non-autochthonous *E. faecium* probiotic strains inclusion in rabbits [21,29,56]. Other authors also noted abundance of microflora in caecum and higher lactobacilli counts after *Lactobacillus* strains application [57]. Outgoing from results of several studies focusing on the molecular profiling of rabbits gut [58–60] and colonization ability of applied enterococcal probiotic strains [33], enterococci were found as predominant microbiota from the phyla *Firmicutes* and they were able also to colonize the caecum and the small intestine (ileum and jejunum) in sufficient counts. The microbiological examinations in fecal samples confirmed the presence of *E. faecium* CCM7420 strain. This strain was able to colonize the digestive tract of rabbits, reaching counts in the range 2.8–6.7 log cycle during the 2 or 3 weeks treatment. These numbers are comparable with the level of autochthonous probiotic strains of several *Enterococcus* spp. after their application in rabbits [35]. Obviously, decreased CCM7420 counts (1.0–3.3 log cycle) were noted in the post-treatment period (3 weeks after strain cessation), but it was still able to persist in the rabbit's intestine. However, the lowest counts in fecal samples were achieved during the experimental application of the lyophilized CCM7420 strain [60], this level is comparable to that one achieved through fresh culture addition of non-autochthonous, bacteriocinogenic and the probiotic *E. faecium* AL41 (CCM8558) strain in rabbits [21]. Important results concerning the spoilage microbiota, including clostridia, coliforms and staphylococci were described in our experiments. Significant reduction of coagulase-positive staphylococci was noted in fecal samples of rabbits, administering fresh culture of CCM7420 ($p < 0.01$; [24]); the *S. aureus* counts were also reduced by 0.5–1.2 log; $p < 0.001$ [25,26]. The CCM7420 strain seems to be useful in rabbits suffering from diarrhea disturbances involving coliforms (reduced by 1.6 log; $p < 0.001$ [26]) and/or clostridia (reduced by 0.5–1.5 log; $p < 0.05$ [25,26]). Similarly to our results, decreased counts of coliforms, *S. aureus* and clostridia were presented in probiotic-treated rabbits [21,23,29]. In rabbits administering the enterocin Ent7420, reduction in coliforms, coagulase-positive staphylococci, including *S. aureus* was noted [25]. Our results are in accordance to those presented by Lauková et al. [21], who tested the effect of nisin on the rabbits gut microbiota and noted reduced counts of the most bacterial species. The gut microbiota is an important constituent in the intestine's mucosal barrier; the increase of the host defense had been already demonstrated by the application of potentially beneficial microorganisms and other natural antimicrobials (bacteriocins, organic acids and plant extracts; [61]).

Changes in bacterial composition—decrease of fecal coliforms, *Pseudomonas*-like sp., *Clostridium*-like sp. and *S. aureus*—during the additives supplementation confirm the antibacterial effect of CCM7420 strain. The inhibitory effect of the bacteriocin-producing and probiotic enterococci and their enterocins on rabbits' intestinal microbiota was already reported in our previous studies [24,25,28,29,52]. Kritas et al. [23] also described the lower frequency of *E. coli* and *C. perfringens* in rabbits treated by probiotic. The dominancy in antimicrobial activity of CCM7420 strain combined with *E. senticosus* can be confirmed by the fact that the *E. senticosus* extract possesses slight or no antimicrobial activity [26]. We supposed that the dietary modulation of the gastrointestinal microbiota by natural antimicrobial substances could result in an enhancement of colonization resistance against potentially pathogenic bacteria.

3.3. *Effect on Eimeria sp. Oocysts*

Coccidiosis is one of the most frequent and prevalent parasitic diseases in rabbit farms. The most markedly and typical symptoms are weight loss, diarrhea (from mild intermittent to severe) with the presence of blood and/or mucus in feces, which leads, through dehydration, to the mortality of animals. The high morbidity and mortality rates among all ages, especially in the young rabbits may be responsible for important economic losses [62]. The oocysts are always present in the intestines of rabbits and they cannot be completely eliminated even by the use of coccidiostat because of the caecotrophy and the symptomless, but potential source of infections of adults. These oocysts are able to cause not only pure eimeriosis after their multiplication and massive infection, but they may also be the cause of multifactorial diseases, when associated with other bacterial or viral infections. Eimeria infections can cause severe disease depending on *Eimeria* species, especially in young animals and the highest incidence of oocysts was usually found around the weaning period [63]. The EU has banned the use of antibiotics as feed additives for growth promotion in animals since 2005 [64]. Today, these antibiotics are replaced with alternative anticoccidials, including prebiotics and probiotics, based on their bactericidal and/or bacteriostatic activities, with immunostimulation and improved growth performance and productivity of the host organism. The experimental application of the CCM7420 strain and its Ent7420 was associated with the reduction of fecal *Eimeria* sp. oocysts. The different treatment period had no impact on oocysts reduction. When the fresh culture of the CCM7420 strain at concentration $\times 10^9$ CFU/mL/g was applied only for 2 weeks to compare its effect with the commonly used probiotic strain *Lactobacillus rhamnosus* GG, the decrease in oocysts counts was observed after probiotic application (8.3×10^1 OPG; oocysts per 1 g of feces) compared to control data (1.5×10^4 OPG), but also compared to the initial counts in experimental group (7.5×10^2 OPG). This reduction effect was maintained until the end of the experiment, also after cessation of the CCM7420 strain. Moreover, at the end of the experiment (at day 42), the difference one order of magnitude in *Eimeria* sp. oocysts was found comparing the control (1.5×10^4 OPG) and the experimental groups (7.2×10^3 OPG; [26]). On the other hand, when CCM7420 was applied with its bacteriocin Ent7420 in rabbits through 3 weeks, after 1 week of their addition, oocysts showed a trend towards a numerical reduction (not significant; [25]) in both experimental groups compared to the control group (Table 1). Surprisingly, at the end of CCM7420 strain addition, increased oocyst counts was observed in this group, while Ent7420 administration was more effective due to a significant reduction in oocysts ($p < 0.05$) at the end of probiotic application (day 21). This finding could be explained by the irregular excretion of oocysts. In the group with Ent7420 addition, a decreasing tendency of oocysts occurrence (not significant) was observed up to the end of enterocin substance application. This could lead to consideration that longer application of CCM7420 strain did not influence more the *Eimeria* sp. oocysts counts in rabbits.

While probiotics are widely used in animals because they improve the growth performance, productivity, health status and stimulate the immunity, studies concerning the protective effect against *Eimeria* sp. are still limited and focused on mainly the avian coccidiosis [65,66]. In rabbits, natural alternatives—prebiotics and herbal extracts—to coccidiostats have been studied [52,67–69]. To the best of our knowledge, only several works demonstrated the anticoccidial effect of beneficial microbes and/or probiotics as well as their antimicrobial products bacteriocins in rabbits [24,25,28,52,53,55,70]. The in vitro effect of four probiotic/bacteriocin-producing strains towards poultry *Eimeria* sp. oocysts was also documented by Strompfová et al. [71] and no differences in the reductive effect of bacteriocin-producing and non-producing strains ($p < 0.05$) were found in this experiment. The in vivo administration of bacteriocin-producing and probiotic strains decreased *Eimeria* sp. oocysts in rabbits. Outgoing from these results, we supposed that anticoccidial effect could be done due to the lactic acid production or by the effect of bacteriocins produced by the mentioned strains. Moreover, the potential protective effect of the CCM7420 strain against zoonotic *Trichinella spiralis* infection was also investigated in the framework of a new therapeutic strategy aimed at using probiotics to control parasitic zoonoses [72], when the authors demonstrated the reduced intensity of *T. spiralis* infection and female fecundity ex vivo and in vitro (about 60%) throughout the CCM7420 administration.

3.4. Effect on Serum Biochemistry

The measurement of biochemical parameters is data mostly used for diagnostic investigations and presents a useful way for controlling the health of animals. However, in some cases, several "components" of the host biochemistry are less specific because of the reparation/compensation ability of healthy tissues/organs and/or metabolic processes.

The tested serum parameters were in the range of normal values defined for these parameters in previous studies with rabbits [73–75], although there are differences in physiological or reference ranges in rabbit serum. During the *E. faecium* CCM7420 application, increased (even though not significantly) concentration of the total protein (TP) was noted in most experiments, and remained stable or higher also three weeks after the strain ceasing (cessation). The highest increase in TP was measured through the fresh CCM7420 culture administration in rabbits (Table 3) [24–26], while the lyophilized (freeze dried form resolved in water and mixed in pellets did not affect the TP level [60]. Application of enterocin Ent7420 as well as the combinative application of CCM7420 with the *E. senticosus* extract also improved the TP concentration [25,26]; similarly as was reported by Fathi et al. [33] and Kalma et al. [51] who observed a slight increase in serum TP after probiotic supplementation in rabbits. The increased level of TP could be explained by better resorption and utilization from the gut; this finding could be also confirmed by higher ADWG. On the other hand, application of non-autochthonous probiotic strains *E. faecium* CCM4231 and AL41 as well as their enterocins did not affect the TP in blood serum [21,29,38]. Blood glucose is an important source of energy for many cells and this is a parameter of the balance between glucose source/availability and utilization. Similarly to TP, higher glucose content was observed during fresh (both 2 and 3 weeks) and lyophilized CCM7420 culture mixed in feed as well as during Ent7420 application. The increased glucose level can be explained by conversion of lactic acid to pyruvate through the gluconeogenesis in the liver. Oppositely, reduced glucose level was observed in the case of the rehydrated–lyophilized CCM7420 strain and its combination with *E. senticosus* ($p < 0.01$). It could be that the glucose concentration was reduced by increased H^+ concentration due to higher organic acid values in the cecum content, which inhibited gluconeogenesis [76]. The increased H^+ (lactate accumulation) in the organism first stimulates physicochemical mineral dissolution by increasing the osteoclast and osteoblast activity (bone resorption) and mostly the Ca^{2+} and Mg^{2+} reabsorption in renal tubules for pH neutralization; it is usually confirmed also with higher serum calcium levels. This hypothesis was confirmed also by us, when the slight increase of calcium content was noted during three weeks CCM7420 fresh culture and its Ent7420 application. The rabbits' blood calcium levels fluctuate widely, dependent upon the level of calcium in their diet and the intestinal absorption as well [77]; this is a difference in the calcium metabolism from other mammals. While Lauková et al. [21] and Szabóová et al. [35] described no influence of probiotic strains on serum glucose, triglycerides and calcium levels, Fathi et al. [33] presented numerical increase of triglycerides associated with dietary probiotic treatment in rabbits, similarly to our achievements [25]. The hypocholesterolemic effect of probiotics in rabbits has been already presented [30,34,51]; surprisingly, our results did not confirm those findings. Moreover, increased/higher cholesterol levels (however, without significant changes) were measured during fresh culture CCM7420 and its combinative administration with *E. senticosus* [26,55] in rabbits; the detected levels were still within the physiological norm.

Table 3. The effect of *Enterococcus faecium* CCM7420 and its enterocin Ent7420 on the serum biochemistry of rabbits.

	E. faecium CCM7420 Strain					Enterocin (Ent)7420
Tested Blood Parameters	**Fresh Culture**	**Fresh Culture**	**Lyophilized Form Resolved in Water**	**Lyophilized Form Mixed into Pellets**	**Fresh Culture + *E. senticosus***	
Reference number of publication	[24]	[25]	[55]	[55]	[26]	[25]
Length of application	14 days	21 days	21 days	21 days	21 days	21 days
Total proteins (g/L)	Increased ($p < 0.05$)	Increased	Unchanged	Unchanged	Increased	Increased ($p < 0.05$)
Total lipids (g/L)	Increased	Increased	Unchanged	Unchanged	Unchanged	Increased
Cholesterol (mmol/L)	Not tested	Not tested	Unchanged	Unchanged	Increased	Not tested
Glucose (mmol/L)	Unchanged	Increased	Increased	Decreased	Decreased	Increased
Calcium (mmol/L)	Unchanged	Increased	Unchanged	Unchanged	Not tested	Increased
Glutathione-peroxidase (GSH-Px; U/mL)	Increased	Decreased	Increased	Increased	Decreased	Decreased
Phagocytic activity (%)	Not tested	Not tested	Increased	Increased	Increased ($p < 0.0001$)	Not tested

The fresh culture of CCM7420 strain was applied into water (at concentration of cells 1×10^9 CFU/mL; dose 500 μL/animal /day); lyophilized (freeze-dried) from rehydrated in water (1×10^9 CFU/mL; dose 500 μL/animal /day) as well as mixed in feed and pelleted (15 g/100 kg feed). Statistical analysis was performed using one-way analysis of variance (ANOVA) with the post hoc Tukey test with the level of significance set at ($p < 0.05$), within experimental groups during each individual experiments.

Exogenous factors such as manipulation, nutritional, weather and temperature changes (mainly hot environmental conditions) often induce physiological oxidative stress, which is avoided by the host's defense system. The host's reaction to stress can be marked mostly by the glutathione-peroxidase (GPx) enzyme activity in blood. In addition, there were no significant differences in GPx activity in blood among experimental groups whereas they were differently affected by CCM7420 (Table 3). Application of CCM7420 strain did not disturb the GPx level; while Ent7420 has a reducing effect on GPx during its application (day 21; $p < 0.05$; [25]). Comparing our previous experimental applications of probiotic strains and their enterocins in rabbits, we suppose that enterocins were more active to protect the host organism; our assumptions were confirmed by reduced GPx levels during Ent7420 and EntM application [25,36]. Outgoing from results we suppose that application of CCM7420 and its bacteriocin Ent7420 did not evoke oxidative stress in the rabbits, similarly to other probiotic strains or enterocins administration [21,29,38].

3.5. Effect on Organic Acids

In rabbits, approximately 40% of digested organic matter of the feed is digested in the caeco-colic segment; so the caecum and the proximal colon are the primary fermenters [78]. The digestion process of nutrients continues in the small intestine by the digestive enzymes of the host, but some components, e.g., plant cell walls and fibers (mainly lignins, cellulose, hemicellulose, pectins) are hydrolyzed by bacterial enzymes into soluble smaller compounds and fermented into the end products: volatile fatty acids (VFA: acetic, propionic and butyric acid), ammonia, intermediary metabolites (lactic, succinic and formic acid) and gas (CO_2, CH_4 and H_2; [79]). The stable microbial fermentation is essential for rabbit health, and only small dietary and environmental changes can lead to increased morbidity and/or mortality via microbial dysbiosis and digestive disturbances. Natural substances applications could prevent those disturbances [12–14]. The concentrations of VFA are usually measured in the cecal content of rabbits [21,28–30,38,78]; in our first experiment we decided to follow the VFA and organic acids concentrations in feces (it was a model experiment with a low number of rabbits in the experimental groups). Application of *E. faecium* CCM7420 to rabbits led to an increase of fecal levels of acetic acid ($p < 0.001$) compared to control animals [24], while other tested organic acids (butyric, succinic and lactic acids) were unaffected with the CCM7420 treatment. Similarly to our results, application of other probiotics (*E. faecium* CCM4231; [28] or bacteriocins (EntM and nisin; [21,38]) in rabbits did not influence the molar proportion of VFA in caecum, while the total VFA production and cecal fermentative activity was increased after *L. plantarum* spray application [29]. Concluded from these results—enhanced enzymatic activity and organic/fatty acid production, better feed conversion ratio and improved jejunal morphology (data shown below) during the CCM7420 strain application, we hypothesized a positive correlation between weight gain and cecal fermentation, improved gut functionality (jejunal morphology) and nutrient absorption. Despite many reports presenting the beneficial results during natural substance application in rabbits, there is a need to extend the existing knowledge and find new possibilities to improve the cecal fermentation and rabbit gut health.

3.6. *Effect on Immunity and Jejunal Morphometry*

Knowledge of the immune response and homeostasis in farm animals represents important information to protect animals from especially bacteria-derived diseases, to improve their health and productivity. The overall organization of the rabbit's digestive immune—lymphoid system is mostly similar to other species, but at the same time it is very special. There are two additional structures identified only in this species, the sacculus rotundus and the vermiform appendix (a place of lymphoid cells differentiation and maturation); they generally act synergistically. The gut microbiota contribute to intestinal homeostasis via inducing the intestinal immune cells and also influence the systemic host immunity. The probiotic consumption shows a beneficial effect in several ways, including intestinal microbiota balance and ability to modulate host innate and specific immune response. Their effect on non-specific immunity was reported as enhanced phagocytosis of pathogenic bacteria and modified cytokine production [80], while the specific way is usually followed through immunoglobulins testing. Nevertheless, the effect of probiotic administration on the immune system of rabbits has been reported on a limited scale [61,81]. Our studies with *E. faecium* CCM7420 alone and in combination with *Eleutherococcus senticosus* demonstrated significant increases in total phagocytic activity (PA) of leukocytes and PA of neutrophils at the end of the treatment period (21 days) and also after three weeks of the ost-treatment period (42 days; $p < 0.0001$; Table 3; [26]). The freeze-dried CCM7420 strain has not influenced the PA in rabbits during its application, either resolved in water or composed in pellets. However, the prolonged effect of CCM7420 strain rehydrated in water was observed at the end of the experiment (42 days; $p < 0.0001$; [55]). Another rabbit studies with non-autochthonous strains *E. faecium* CCM4231 (ruminal isolate) and AL41 (CCM8558; isolate from animal waste) also showed stimulation of non-specific immune reaction in rabbits; the significant increase of PA ($p < 0.001$) was noted in both experiments during the treatment and increased several weeks after the strain's cessation. Fathi et al. [33] also represented improved cell-mediated immunity adding 400 g probiotic/t feed in rabbits´ diet. Contrary to results reported above, Wang et al. [57] noted no probiotic influence on the number of mast cells in duodenum and jejunum, but increased the number of mast cells in caecum and also, increased IgG and IgM in serum. During enterocin Ent7420 administration in rabbits, the prolonged immuno-stimulative effect was observed, which was demonstrated by a significant increase of PA in the experimental group ($p < 0.05$) compared also to the control data [82]. The same immunomoderate influence was noted during experimental applications of enterocins produced by strains CCM4231 and AL41 (Ent4231 and EntM/EntAL41; [21,28]) as well as in the case of nisin feed inclusion in rabbits [38]. We expected that enterocins are able to stimulate the immune system through the gut microbiota modulation on behalf of lactic acid bacteria (LAB) and via supporting/improving the GALT by the stimulation of the IgA system. This fact could explain the "timeshift" of the enterocins influence compared to the probiotic strain application, since probiotic bacteria begins to act and multiply immediately, while bacteriocins firstly modulate the environment on behalf of beneficial microbes (LAB) inhibiting other bacterial species in the gut. On the other hand, the prolonged or maintaining immunostimulative effect (even at 21 days after ceasing the administration of enterocins) should be also explained by the adopting of animals on them. More studies testing the effect of the CCM7420 strain as well as its enterocin Ent7420 on the immune response, containing PA and other parameters of innate immunity in rabbits are necessary. Moreover, we would like to focus also on the intestinal immunity, mainly on the IgA level.

It is well known, that probiotic supplementation improves not only the growth rate and enhances the efficiency of feed conversion but also may positively influence the health status via enhancing gut health in rabbits [61]. Gut health, including microbial and immunological stability, is often influenced by exogenous factors (dietary changes, stress from manipulation, transfer, climate changes, etc.), mainly around the weaning period. Therefore, the alternative strategies are required to improve the animal's health. The stable or improved intestinal environment (gut microflora, mucosal immunity, epithelial morphology and function) directly influences the health status and growth performance of animals due to better nutrient absorption in the gut [61,83]. The effect of natural feed additives, including probiotics on the intestinal histomorphology in animals is often presented, mostly in chicken and pigs. However, studies reported changes of morphometric parameters in the small intestine of rabbits during probiotic application are limited [31,82–84]. Our results showed that the surface area and villi height: crypt depth (VH:CD) ratio also increased throughout the *E. faecium* CCM7420 strain administration [83]. These results could support the hypothesis about higher weight gain, better feed conversion and nutrient utilization due to enlargement of the absorption surface and improved morphometry parameters. Similarly to these results, increased surface area, VH, VH:CD and decreased CD was observed after Ent2019 addition in rabbits. To the best of our current knowledge, our team is the first that deals with experiments regarding the morphological changes in rabbits during enterocins administration (Ent2019 and EntM; [80,85]). This knowledge leads us to a more detailed study of these physiological changes in the rabbit's digestive tract, which will not only allow us to extend our knowledge, but also gain new information to understand the complexity of physiological, microbiological and immunological processes in the host organism.

3.7. *Effect on Meat—Nutrient Content and Physicochemical Properties*

Rabbit meat is greatly valued for its high nutritional and dietary quality, especially its low amounts of cholesterol, fat and sodium and high content of polyunsaturated fatty acids (PUFA), potassium, phosphorus and magnesium [85,86]; for these reasons it is recommended mainly for children, pregnant women and patients with cardiovascular illnesses. Previously there have been many studies/reviews concerning rabbit meat, including its production, quality, physicochemical properties and composition [1,2]. A lot of them dealing with the effects of dietary supplementation with functional compounds, mainly probiotics, prebiotics, fatty and organic acids, vitamins, selenium and antioxidants and their combinations on rabbit carcass quality [12–14,31,33,40,87–89]. In general, there is a lack of studies testing the beneficial microbiota and/or their antimicrobial substances (bacteriocins) on rabbit meat quality and composition, including fatty acids, amino acids and minerals [31,41–45]. Moreover, to the best of our knowledge we presented the first reports about the effect of beneficial/probiotic strains as well as bacteriocins on the minerals and amino acids of rabbit meat [39,40,42] and also the effect of bacteriocins on the fatty acid content of rabbit meat [41–43]. All presented results concerning the physicochemical properties—pH, protein, fat, ash and water content, energy value, lightness and color of rabbit meat—showed that probiotic administration in rabbits had no negative effect on the rabbit meat quality. On the other hand, Fathi et al. [33] presented a significant effect of probiotic supplementation on moisture, dry matter, organic matter and ash content. Only limited information about rabbit minerals are available, in response to natural feed additives. Our result showed increased iron content (p = 0.0011) in treated groups with freeze-dried CCM7420 strain (both, rehydrated in water and enriched in feed pellets), in contrary to other findings with reduced iron content during microbial fermented feed utilization with *Lactobacillus plantarum* and *Pediococcus acidilactici* [90] and after enterocin administration [42]. Although the pH of the luminal content was not measured, we hypothesized a more acidic environment in the gut due to the previous results of higher lactic acid production and lower pH in the caecum during CCM7420 strain administration in rabbits [48]. This acidic environment can enhance the ionization of minerals, which in turn results in passive diffusion [91] and could be one alternative explanation of the higher iron absorption from the gut. Another hypothesis could be the larger absorption surface due to enterocyte proliferation, which is confirmed by improved morphometry parameters—villus height, crypt depth and villus height:crypt depth ratio—also recorded during our previous experiments with CCM7420 administration [82] and lantibiotic–nisin application to rabbits [38]. The enlargement of the luminal surface could ensure better mineral absorption and their inclusion to rabbit meat. On the other hand, significantly decreased concentrations of copper (p = 0.0004) and calcium (p < 0.0001) were noted. Similarly to these results, the copper concentrations in rabbit meat also reduced during the enterocin M addition to rabbits, however, not significant but only numerically changes were recorded. Similarly to us, lower concentrations of copper, zinc and manganese was found by Shah et al. [90] after probiotic supplementation. Copper is an essential trace element, performing important biochemical functions; its level in rabbit meat varies widely [92,93]. Regarding the higher iron concentration, we hypothesized an iron competitive influence on the copper intestinal absorption and its lower deposition to meat. The calcium metabolism in rabbits is very unique, widely fluctuates and its intestinal absorption is very vitamin D independent, in contrast to most mammals [77]. Despite the generally known fact that probiotics can increase the organic and short chain fatty acids in caecum, which also stimulate the minerals ionization and diffusion through the intestine, we noted decreased calcium concentration, although, still in the range presented in the literature [86,92]. Inferring from achieved results, we assumed no adverse effect of the CCM7420 strain on the meat characteristics; in addition, it could enhance the mineral quality of meat and also increased its value to the functional food level.

4. Conclusions

The strain *E. faecium* CCM7420 in different application forms (fresh culture at concentration 1×10^9 CFU/mL of cells in a dose 500 μL/animal /day applied into drinking water) was lyophilized (freeze dried) from rehydrated water (1×10^9 CFU/mL; dose 500 μL/animal/day) as well as mixed in feed and pelleted (15 g/100 kg feed), either alone and in combination with *Eleutherococcus senticosus* and its enterocin Ent7420 (50 μL/animal/day applied into drinking water) were tested in rabbits. During these experiments, the following effects of the strain and its enterocin were observed: improved average daily weight gain and feed conversion ratio, good colonization ability of the tested strain with maximum counts in the first 2–3 weeks of application, increased lactic acid bacteria and reduced coagulase-positive staphylococci including *S. aureus*, coliforms and clostridia population as well as the *Eimeria* sp. oocysts counts in the rabbit's gut. Improved biochemical blood parameters (total proteins, glucose and triglycerides) have been noted during the CCM7420 strain application; however, the glutathione-peroxidase level was not disturbed and oxidative stress was not evoked through the additives application. Another interesting finding was the significant stimulation of blood phagocytic activity and also the improved morphometry parameters (enlargement of the absorption surface in jejunum and higher villi height: crypt depth (VH:CD) ratio). The physicochemical properties of rabbit meat were not negatively affected by the CCM7420 strain, while the meat iron content significantly increased during its application, which improved the rabbit meat quality. It could be also emphasized that knowing the probiotic properties and the ability of *Enterococcus faecium* CCM7420 to produce enterocin Ent7420 with an antimicrobial effect is of great interest mainly in the case of several disease/pathologies, such as epizootic rabbit enteropathy, which are difficult to prevent and combat because their etiology is not known and there is no vaccine. This strain is the main component of the Prorabbit probiotic preparation, which is often used in Slovak rabbit farms (at dosage 1–2 g/animal/day for 21 days as prevention and 3 g/animal/day with a therapeutic effect; resolved in water or mixed into feed). Moreover, to the best of our current knowledge, our team is the first that deals with experiments regarding the morphological changes in the jejunum of rabbits during enterocins administration; the first reports regarding the effect of beneficial/probiotic strains and bacteriocins on the mineral and amino acid concentrations as well as the effect of bacteriocins on fatty acid content of rabbit meat were also published by our team.

Author Contributions: Conceptualization, M.P.S. and A.L.; data curation, M.P.S. and L'.C.; funding acquisition, A.L.; investigation, M.P.S., A.L. and L'.C.; project administration, A.L.; resources, A.L.; validation, A.L.; visualization, M.P.S.; writing—original draft preparation, M.P.S.; writing—review and editing, A.L., M.P.S. All authors have read and agreed to the published version of the manuscript.

Funding: This work was financially supported by the national Slovak Grant Agency VEGA project 2/0006/17.

Acknowledgments: This work was supported by the project VEGA 2/0006/17. All care and experimental procedures involving animals followed the guidelines stated in the Guide for the Care and Use of Laboratory Animals approved by the State Slovak Veterinary and Food Administration and the Ethics Committees of both institutions (permission code: SK CH 17016 and SK U 18016). None of the authors of this paper has a financial or personal relationship with other people or organizations that could inappropriately influence or bias the content of the paper. We would like to thank Mrs. M. Bodnárová and Mr. P. Jerga for their skillful technical assistance with sample collection and processing. We are grateful to A. Kopčáková for the *E. faecium* CCM7420 strain sequencing, L'. Ondruška, V. Párkányi and R. Jurčík, from the National Agricultural and Food Centre (NAFC) in Nitra for blood sampling and Mr. J. Pecho for animals slaughtering, to Z. Formelová and M. Chrenková (NAFC) for meat samples processing and meat quality properties analyses. We also thank to V. Strompfová for blood biochemistry analyses, K. Čobanová and Š. Faix for glutathione-peroxidase activity measuring, I. Plachá for phagocytic activity determination, S. Gancarčíková (University of Veterinary Medicine and Pharmacy, Laboratory of Gnotobiology, Košice) for organic acids measurement, Z. Vasilková (Parasitological Institute of SAS) for *Eimeria* sp. oocysts detection, R. Žitňan for intestinal morphometry testing.

Conflicts of Interest: The authors declare no conflict of interest. The funders had no role in the design of the study; in the collection, analyses, or interpretation of data; in the writing of the manuscript, or in the decision to publish the results.

References

1. Dalle Zotte, A.; Szendrő, Z. The role of rabbit meat as functional food. *Meat Sci.* **2011**, *88*, 319–331. [CrossRef]
2. Cullere, M.; Dalle Zotte, A. Rabbit meat production and consumption: State of knowledge and future perspectives. *Meat Sci.* **2018**, *143*, 137–146. [CrossRef]
3. Basavaraj, M.; Nagabhushana, V.; Prakash, N.; Appannavar, M.M.; Prashanth, W.; Mallikarjunappa, S. Effect of dietary supplementation of curcuma longaon the biochemical profile and meat characteristics of broiler rabbits under summer stress. *Vet. World* **2011**, *4*, 15–18. [CrossRef]
4. Carabaño, R.; Badiola, I.; Licois, D.; Gidenne, T. The digestive ecosystem and its control through nutritional or feeding strategies. In *Recent Advances in Rabbit Sciences*, 1st ed.; Maertens, L., Coudert, P., Eds.; ILVO: Merehleke, Belgium, 2006; pp. 221–227.
5. Marlier, D.; Dewrée, R.; Lassence, C.; Licois, D.; Mainil, J.; Coudert, P.; Meulemans, L.; Ducatelle, R.; Vindevogel, H. Infectious agents associated with epizootic rabbit enteropathy: Isolation and attempts to reproduce the syndrome. *Vet. J.* **2006**, *172*, 493–500. [CrossRef] [PubMed]
6. Salyers, A.A.; Gupta, A.; Wang, Y.P. Human intestinal bacteria as reservoirs for antibiotic resistance genes. *Trends Microbiol.* **2004**, *12*, 412–416. [CrossRef] [PubMed]
7. European Commission. *IP/03/1058 Council and Parliament Prohibit Antibiotics as Growth Promoters: Commissioner Byrne Welcomes Adoption of Regulation on Feed Additives*; European Commission: Brussels, Belgium, 2003.
8. Barug, D.; de Jong, J.; Kies, A.K.; Verstegen, M.W.A. *Antimicrobial Growth Promoters. Where Do We Go from Here?* 1st ed.; Wageningen Academic Publishers: Wageningen, The Netherlands, 2006; pp. 1–196.

9. Falcão-e-Cunha, L.; Castro-Solla, L.; Maertens, L.; Marounek, M.; Pinheiro, V.; Freire, J.; Mourão, J.L. Alternatives to antibiotic growth promoters in rabbit feeding: A review. *World Rabbit Sci.* **2007**, *15*, 127–140. [CrossRef]
10. European Commission. *Regulation EC/1831/2003 of the European Parliament and of the Council on Additives for Use in Animal Nutrition*; European Commission: Brussels, Belgium, 2003.
11. European Commission. *Regulation EC 1924/2006 of the European Parliament and of the Council on Nutrition and Health Claims Made on Foods*; European Commission: Brussels, Belgium, 2006.
12. Papatsiros, V.G.; Christodoulopoulos, G. The use of organic acids in rabbit farming. *Online J. Anim. Feed Res.* **2011**, *1*, 434–438.
13. Dalle Zotte, A.; Celia, C.; Szendrő, Z. Herbs and spices inclusion as feedstuff or additive in growing rabbit diet and as additive in rabbit meat: A review. *Livest. Sci.* **2016**, *189*, 82–90. [CrossRef]
14. Kalma, R.P.; Patel, V.K.; Joshi, A.; Umatiya, R.V.; Parmar, K.N.; Damor, S.V.; Chauhan, H.D.; Srivastava, A.K.; Sharma, H.A. Probiotic supplementation in rabbit: A review. *Int. J. Agric. Sci.* **2016**, *8*, 2811–2815.
15. Sharma, K.G.; Vidyarthi, K.V.; Archana, K.; Zuyie, R. Probiotic supplementation in the diet of rabbits—A Review. *Livest. Res. Int.* **2016**, *4*, 1–10.
16. Canzi, E.; Zanchi, R.; Camaschella, P.; Cresci, A.; Greppi, G.F.; Orpianesi, C.; Serrantoni, M.; Ferrari, A. Modulation by lactic-acid bacteria of the intestinal ecosystem and plasma cholesterol in rabbits fed a casein diet. *Nutr. Res.* **2000**, *20*, 1329–1340. [CrossRef]
17. Amber, K.H.; Yakout, H.M.; Hamed Rawya, S. Effect of feedings diets containing Yucca extract or probiotic on growth, digestibility, nitrogen balance and caecal microbial activity of growing New Zealand White rabbits. In Proceedings of the 8th World Rabbit Congress, Puebla, Mexico, 7–10 September 2004; pp. 737–745.
18. Trocino, A.; Xiccato, G.; Carraro, L.; Jiménez, G. Effect of diet supplementation with Toyocerin® (*Bacillus cereus* var. *Toyoi*) on performance and health of growing rabbits. *World Rabbit Sci.* **2005**, *13*, 17–28. [CrossRef]
19. Matusevičius, P.; Ašmenskaitė, L.; Žilinskienė, A.; Gugolek, A.; Lorek, M.O.; Hartman, A. Effect of probiotic Bioplus 2B® on performance of growing rabbits. *Vet. Zootech* **2006**, *36*, 54–59.
20. Lauková, A.; Strompfová, V.; Skřivanová, V.; Volek, Z.; Jindřichová, E.; Marounek, M. Bacteriocin producing strain of *Enterococcus faecium* EK13 with probiotic character and its application in the digestive tract of rabbits. *Biol. Bratisl.* **2006**, *61*, 779–782. [CrossRef]
21. Lauková, A.; Chrastinová, L'.; Pogány Simonová, M.; Strompfová, V.; Plachá, I.; Čobanová, K.; Formelová, Z.; Chrenková, M.; Ondruška, L'. *Enterococcus faecium* AL 41: Its enterocin M and their beneficial use in rabbits husbandry. *Probiotics Antimicrob. Proteins* **2012**, *4*, 243–249. [CrossRef]
22. Lauková, A.; Pogány Simonová, M.; Chrastinová, L'.; Kandričáková, A.; Ščerbová, J.; Plachá, I.; Čobanová, K.; Formelová, Z.; Ondruška, L'.; Štrkolcová, G.; et al. Beneficial effect of bacteriocin-producing strain *Enterococcus durans* ED 26E/7 in model experiment using broiler rabbits. *Czech J. Anim. Sci.* **2017**, *62* (Suppl. 4), 168–177.
23. Kritas, S.K.; Petridou, E.I.; Fortomaris, P.; Tzika, E.; Arsenos, G.; Koptopoulos, G. The effect of probiotics on microbiology, health and performance of fattening rabbits. *Asian Australas. J. Anim. Sci.* **2008**, *21*, 1312–1317. [CrossRef]
24. Simonová, M.; Marciňáková, M.; Strompfová, V.; Čobanová, K.; Gancarčíková, S.; Vasilková, Z.; Lauková, A. Effect of probiotics *Lactobacillus rhamnosus* GG and new isolate *Enterococcus faecium* EF2019 (CCM 7420) on growth, blood parameters, microbiota and coccidia oocysts excretion in rabbits. *Int. J. Prob. Preb.* **2008**, *3*, 7–14.
25. Pogány Simonová, M.; Lauková, A.; Chrastinová, L'.; Strompfová, V.; Faix, Š.; Vasilková, Z.; Ondruška, L'.; Jurčík, R.; Rafay, J. *Enterococcus faecium* CCM7420, bacteriocin PPB CCM7420 and their effect in thedigestive tract of rabbits. *Czech J. Anim. Sci.* **2009**, *54*, 376–386. [CrossRef]
26. Pogány Simonová, M.; Lauková, A.; Chrastinová, L'.; Plachá, I.; Čobanová, K.; Strompfová, V.; Formelová, Z.; Chrenková, M. Combined administration of bacteriocin-producing, probiotic strain *Enterococcus faecium* CCM7420 with *Eleutherococcus senticosus* and their effect in rabbits. *Pol. J. Vet. Sci.* **2013**, *8*, 730–734.
27. Pogány Simonová, M.; Chrastinová, L'.; Kandričáková, A.; Kubašová, I.; Formelová, Z.; Chrenková, M.; Miltko, R.; Belzecki, G.; Strompfová, V.; Lauková, A. Enterocin M and sage supplementation in post-weaning rabbits: Effects on growth performance, caecal microbiota, fermentation and enzymatic activity. *Probiotics Antimicrob. Proteins* **2020**, *12*, 732–739. [CrossRef] [PubMed]

28. Szabóová, R.; Lauková, A.; Chrastinová, L'.; Pogány Simonová, M.; Strompfová, V.; Plachá, I.; Čobanová, K.; Vasilková, Z.; Chrenková, M. Enterocin 4231 produced by *Enterococcus faecium* CCM 4231 and its use in rabbits. *Acta Vet. Beogr.* **2011**, *61*, 523–529. [CrossRef]
29. Bovera, F.; Iannaccone, F.; Mastellone, V.; Nizza, S.; Lestingi, A.; De Martino, L.; Lombardi, P.; Mallardo, K.; Ferrara, M.; Nizza, A. Effect of spray application of *Lactobacillus plantarum* on in vivo performance, caecal fermentations and haematological traits of suckling rabbits. *Ital. J. Anim. Sci.* **2012**, *11*, 145–149. [CrossRef]
30. Oso, A.O.; Idowu, O.M.O.; Haastrup, A.S.; Ajibade, A.J.; Olowonefa, K.O.; Aluko, A.O.; Ogunade, I.M.; Osho, S.O.; Bamgbose, A.E. Growth performance, apparent nutrient digestibility, caecal fermentation, ileal morphology and caecal microflora of growing rabbits fed diets containing probiotics and prebiotics. *Livest. Sci.* **2013**, *157*, 184–190. [CrossRef]
31. Bhatt, R.S.; Agrawal, A.R.; Sahoo, A. Effect of probiotic supplementation on growth performance, nutrient utilization and carcass characteristics of growing Chinchilla rabbits. *J. Appl. Anim. Res.* **2017**, *45*, 304–309. [CrossRef]
32. Cunha, S.; Mendes, Â.; Rego, D.; Meireles, D.; Fernandes, R.; Carvalho, A.; Martins da Costa, P. Effect of competitive exclusion in rabbits using an autochtonous probiotic. *World Rabbit Sci.* **2017**, *25*, 123–134. [CrossRef]
33. Fathi, M.; Abdelsalam, M.; Al-Homidan, I.; Ebeid, T.; El-Zarei, M.; Abou-Emera, O. Effect of probiotic supplementation and genotype on growth performance, carcass traits, hematological parameters and imunity of growing rabbits under hot environmental conditions. *Anim. Sci. J.* **2017**, *88*, 1644–1650. [CrossRef]
34. Phuoc, T.L.; Jamikorn, U. Effects of probiotic supplement (*Bacillus subtilis* and *Lactobacillus acidophilus*) on feed efficiency, growth performance, and microbial population of weaning rabbits. *Asian Austral. J. Anim. Sci.* **2017**, *30*, 198–205. [CrossRef]
35. Szabóová, R.; Chrastinová, L'.; Lauková, A.; Haviarová, M.; Simonová, M.; Strompfová, V.; Faix, Š.; Vasilková, Z.; Chrenková, M.; Plachá, I.; et al. Bacteriocin - producing strain *Enterococcus faecium* CCM4231 and its use in rabbits. *Int. J. Prob. Preb.* **2008**, *3*, 77–82.
36. Pogány Simonová, M.; Lauková, A.; Plachá, I.; Čobanová, K.; Strompfová, V.; Szabóová, R.; Chrastinová, L'. Can enterocins affect phagocytosis and glutathione peroxidase in rabbits? *Cent. Eur. J. Biol.* **2013**, *8*, 730–734.
37. Pogány Simonová, M.; Chrastinová, L'.; Kandričáková, A.; Gancarčíková, S.; Bino, E.; Plachá, I.; Ščerbová, J.; Strompfová, V.; Žitňan, R.; Lauková, A. Can have enterocin M in combination with sage beneficial effect on microbiota, blood biochemistry, phagocytic activity and jejunal morphometry in broiler rabbits? *Anim.* **2020**, *10*, 115. [CrossRef]
38. Lauková, A.; Chrastinová, L'.; Plachá, I.; Kandričáková, A.; Szabóová, R.; Strompfová, V.; Chrenková, M.; Čobanová, K.; Žitňan, R. Beneficial effect of lantibiotic nisin in rabbit husbandry. *Probiotics Antimicrob. Proteins* **2014**, *6*, 41–46. [CrossRef]
39. Pogány Simonová, M.; Szabóová, R.; Haviarová, M.; Chrastinová, L'.; Mojto, J.; Strompfová, V.; Lauková, A.; Rafay, J. Quality of rabbit meat after application of bacteriocinogenic and probiotic strain *Enterococcus faecium* CCM 4231 in rabbits. *Int. J. Prob. Preb.* **2009**, *4* (Suppl. 1), 1–6.
40. Pogány Simonová, M.; Chrastinová, L'.; Lauková, A. Dietary supplementation of a bacteriocinogenic and probiotic strain of *Enterococcus faecium* CCM7420 and its effect on the mineral content and quality of *Musculus longissimus dorsi* in rabbits. *Anim. Prod. Sci.* **2016**, *56* (Suppl. 12), 2140–2145.
41. Pogány Simonová, M.; Chrastinová, L'.; Lauková, A. Lantibiotic nisin applied in broiler rabbits and its effect on the growth performance and carcass quality. *Probiotics Antimicrob. Proteins* **2019**, *11*, 1414–1417. [CrossRef]
42. Pogány Simonová, M.; Chrastinová, L'.; Chrenková, M.; Formelová, Z.; Kandričáková, A.; Bino, E.; Lauková, A. Benefits of enterocin M and sage combination on the physico-chemical traits, fatty acid, amino acid, and mineral content of rabbit meat. *Probiotics Antimicrob. Proteins* **2019**. [CrossRef]
43. Chrastinová, L'.; Lauková, A.; Chrenková, M.; Poláčiková, M.; Formelová, Z.; Kandričáková, A.; Glatzová, E.; Ščerbová, J.; Bučko, O.; Ondruška, L'.; et al. Effects of enterocin M and durancin ED26E/7 substances applied in drinking water on the selected carcass characteristics and meat quality of broiler rabbits. *Arch. Zootech.* **2018**, *21*, 5–17.
44. EFSA Panel on Additives and Products or Substances used in Animal Feed (EFSA FEEDAP PANEL). Safety and efficacy of Probiotic Lactina® (*Enterococcus faecium* NBIMCC8270, *Lactobacillus acidophilus* NBIMCC 8242, *Lactobacillus helveticus* NBIMCC 8269, *Lactobacillus delrueckei* ssp. *lactis* NBIMCC 8250, *Lactobacillus delbrueckei* ssp. *bulgaricus* NBIMCC 8244 and *Streptococcus thermophilus* NBIMCC 8253) as a feed additive for chickens for fattening and suckling and weaned rabbits. *EFSA J.* **2019**, *17*, 5646.

45. Matusevičius, P.; Bartkeviciute, Z.; Cernauskenie, J.; Kozłowski, K.; Jeroch, H. Effect of probiotic preparation "ToyoCerin (R)" and phytobiotic preparation "Cuxarom Spicemaster" on growing rabbits. *Arch. Geflügelkunde* **2011**, *75*, 67–71.
46. Simonová, M.; Lauková, A. Isolation of faecal *Enterococcus faecium* strains from rabbits and their sensitivity to antibiotics and ability to bacteriocin production. *Bull. Vet. Inst. Pulawy* **2004**, *48*, 383–386.
47. Simonová, M.; Lauková, A. Bacteriocin activity of enterococci from rabbits. *Vet. Res. Comm.* **2007**, *31*, 143–152. [CrossRef]
48. Simonová, M. Probiotic and Bacteriocin-Producing Bacteria and Their Effect on Physiology of Digestion in Rabbits. Ph.D. Thesis, Institute of Animal Physiology, Slovak Academy of Sciences, Košice, Poland, 2006.
49. Simonová, M.; Lauková, A.; Sirotek, K. (2008b): Lipolytic activity of potential probiotic enterococci and additive staphylococci. *Acta Vet. Brno* **2008**, *77*, 575–580. [CrossRef]
50. El Dimerdash, M.Z.; Dalia, M.H.; Hanan, F.A.; Doaa, S.A. Studies on the effect of some probiotics in rabbits. *Suez Canal Vet. Med. J.* **2011**, *16*, 151–168.
51. Kalma, R.P.; Chauhan, H.D.; Shrivastava, A.K.; Pawar, M.M. Growth and blood profile of broiler rabbits on probiotics supplementation. *Ind. J. Small Rum.* **2018**, *24*, 66–69. [CrossRef]
52. Szabóová, R.; Chrastinová, L.; Lauková, A.; Strompfová, V.; Pogány Simonová, M.; Vasilková, Z.; Plachá, I.; Čobanová, K.; Chrenková, M. Effect of combinative administration of bacteriocin-producing and probiotic strain *Enterococcus faecium* CCM 4231 and sage plant extract on rabbits. *Afr. J. Microbiol. Res.* **2012**, *6*, 4868–4873. [CrossRef]
53. Lauková, A.; Chrastinová, L'.; Plachá, I.; Szabóová, R.; Kandričáková, A.; Pogány Simonová, M.; Formelová, Z.; Ondruška, L'.; Goldová, M.; Chrenková, M.; et al. Enterocin 55 produced by non rabbit-derived strain *Enterococcus faecium* EF55 in relation with microbiota and selected parameters in broiler rabbits. *Int. J. Agric. Environ. Res.* **2017**, *3*, 45–52.
54. Lauková, A.; Pogány Simonová, M.; Chrastinová, L'.; Gancarčíková, S.; Kandričáková, A.; Plachá, I.; Chrenková, M.; Formelová, Z.; Ondruška, L'.; Ščerbová, J.; et al. Assessment of Lantibiotic type bacteriocin—Gallidermin application in model experiment with broiler rabbits. *Int. J. Anim. Sci.* **2018**, *2*, 1028.
55. Pogány Simonová, M.; Lauková, A.; Chrastinová, L'.; Strompfová, V.; Plachá, I.; Szabóová, R.; Chrenková, M.; Čobanová, K.; Goldová, M.; Žitňan, R. Benefity of probiotic and bacteriocin-producing strain Enterococcus faecium EF2019 (CCM7420) for rabbits health (In Slovak). In Proceedings of the 12th Conference of New Ways in Intensive and Hobbies Rabbits Husbandries, Praha, Czech Republic, 6 November 2013; pp. 37–43.
56. Benato, L.; Hastie, P.; O'Shaughnessy, P.; Murray, J.A.; Meredith, A. Effects of probiotic Enterococcus faecium and Saccharomyces cerevisiae on the faecal microflora of pet rabbits. *J. Small Anim. Pr.* **2014**, *55*, 442–446. [CrossRef]
57. Wang, C.; Zhu, Y.; Li, F.; Huang, L. The effect of *Lactobacillus* isolates on growth performance, immune response, intestinal bacterial community composition of growing Rex rabbits. *Anim. Physiol. Anim. Nutr.* **2017**, *101*, e1–e13. [CrossRef]
58. Linaje, R.; Coloma, M.D.; Pérez-Martínez, G.; Zúñiga, M. Characterization of faecal enterococci from rabbits for the selection of probiotic strains. *J. Appl. Microbiol.* **2014**, *96*, 761–771. [CrossRef]
59. Abecia, L.; Fondevila, M.; Balcells, J.; Edwards, J.E.; Newbold, J.C.; McEwan, N.R. Molecular profiling of bacterial species in the rabbit cecum. *FEMS Microbiol. Lett.* **2005**, *244*, 111–115. [CrossRef] [PubMed]
60. Simonová, M.; Lauková, A.; Štyriak, I. Enterococci from rabbits—potential feed additives. *Czech J. Anim. Sci.* **2005**, *50*, 416–421. [CrossRef]
61. Fortune-Lamothe, L.; Boullier, S. A review on the interactions between gutmicroflora and digestive mucosal imunity. Possible ways to improve the health of rabbits. *Livest. Sci.* **2007**, *107*, 1–18. [CrossRef]
62. Pakandl, M. Coccidia of rabbit: A review. *Folia Parasitol.* **2009**, *56*, 153–166. [CrossRef]
63. Zita, L.; Tůmová, E.; Skřivanová, V.; Ledvinka, Z. The effect of weaning age on performance and nutrient digestibility of broiler rabbits. *Czech J. Anim. Sci.* **2007**, *52*, 341–347. [CrossRef]
64. EC (2005) European Union Commission. *Ban on Antibiotics as Growth Promoters in Animal Feed Enters into Effect: Regulation 1831/2003/EC on Additives for Use in Animal Nutrition, Replacing Directive 700/524//333c on Additives in Feedstuffs*; European Union Commission: Brussels, Belgium, 2005.
65. Travers, M.-A.; Florent, I.; Kohl, L.; Grellier, P. Probiotics for the control of parasites: An overview. *J. Parasitol. Res.* **2011**, *2011*, 610769. [CrossRef]

66. Markowiak, P.; Śliżewska, K. The role of probiotics, prebiotics and synbiotics in animal nutrition. *Gut Pathog.* **2018**, *10*, 21. [CrossRef]
67. Kowalska, D.; Bielański, P.; Nosal, P.; Kowal, J. Natural alternatives to coccidiostats in rabbit nutrition. *Ann. Anim. Sci.* **2012**, *12*, 561–574. [CrossRef]
68. Nosal, P.; Kowalska, D.; Bielanski, P.; Kowal, J.; Kornas, S. Herbal formulations as feed additives in the course of rabbit subclinical coccidiosis. *Ann. Parasitol.* **2014**, *60*, 65–69.
69. El-Ashram, S.A.; Aboelhadid, S.M.; Abdel-Kafy, E.-S.M.; Hashem, S.A.; Mahrous, L.N.; Farghly, E.M.; Moawad, U.K.; Kamel, A.A. Prophylactic and therapeutic efficacy of prebiotic supplementation against intestinal coccidiosis in rabbits. *Animals* **2019**, *9*, 965. [CrossRef]
70. Szabóová, R.; Lauková, A.; Chrastinová, Ľ.; Strompfová, V.; Pogány Simonová, M.; Vasilková, Z.; Čobanová, K.; Plachá, I.; Chrenková, M. Effect of combined administration of enterocin 4231 and sage in rabbits. *Pol. J. Vet. Sci.* **2011**, *14*, 359–366. [CrossRef] [PubMed]
71. Strompfová, V.; Lauková, A.; Marciňáková, M.; Vasilková, Z. Testing of probiotic and bacteriocin-producing lactic acid bacteria towards *Eimeria* sp. *Pol. J. Vet. Sci.* **2010**, *13*, 389–391. [PubMed]
72. Bucková, B.; Hurníková, Z.; Lauková, A.; Revajová, V.; Dvorožňáková, E. The anti-parasitic effect of probiotic bacteria via limiting the fecundity of *Trichinella spiralis* female adults. *Helminthol.* **2018**, *55*, 102–111. [CrossRef] [PubMed]
73. Jenkins, J.F. Clinical Pathology. In *Manual of Rabbits Medicine and Surgery*, 2nd ed.; Meredith, A., Flecknell, P., Eds.; BSAVA: Gloucester, UK, 2006.
74. Martinec, M.; Härtlová, H.; Chodová, D.; Tůmová, E.; Fučíková, A. Selected haematological and biochemical indicators in different breeds of rabbits. *Acta Vet. Brno* **2012**, *81*, 371–375. [CrossRef]
75. Özkan, C.; Kaya, A.; Akgül, Y. Normal values of haematological and some biochemical parameters in serum and urine of New Zealand rabbits. *World Rabbit Sci.* **2012**, *20*, 253–259. [CrossRef]
76. Illes, R.A.; Cohen, R.D.; Rist, A.H.; Baron, P.G. The mechanism of inhibition by acidosis of gluconeogenesis from lactate in rat liver. *Biochem. J.* **1977**, *164*, 185–191. [CrossRef] [PubMed]
77. Redrobe, S. Calcium metabolism in rabbits. In *Seminars in Avian and Exotic Pet Medicine*; WB Saunders: Philadelphia, PA, USA, 2002; Volume 11, pp. 94–101.
78. Gidenne, T. Estimation of volattile fatty acids and of their energetic supply in the rabbit caecum: Effect of the dietary fibre level. In Proceedings of the Conference Vleme Journees de la Recherche Cunicole, Paris, France, 6–7 December 1994; pp. 293–299.
79. Combes, S.; Fortune-Lamothe, L.; Cauqui, L.; Gidenne, T. Engineering the rabbit digestive ecosystem to improve digestive health and efficacy. *Animal* **2013**, *7*, 1429–1439. [CrossRef]
80. Maldonado, G.C.; Lemme-Dumit, J.M.; Thieblemont, N.; Carmuega, E.; Weill, R.; Perdigón, G. Stimulation of innate immune cells induced by probiotics: Participation of toll-like receptors. *J. Clin. Cell. Immunol.* **2015**, *6*, 1000283.
81. Deptuła, W.; Niedźwiedzka-Rystwej, P.; Śliwa, J.; Kaczmarczyk, M.; Tokarz-Deptuła, B.; Hukowska-Szematowicz, B.; Pawlikowska, M. Values of selected immune indices in healthy rabbits. *Centr. Eur. J. Immunol.* **2008**, *33*, 190–192.
82. Pogány Simonová, M.; Lauková, A.; Žitňan, R.; Chrastinová, Ľ. Effect of rabbit origin enterocin-producing strain *Enterococcus faecium* CCM7420 application on growth performance and gut morphometry in rabbits. *Czech J. Anim. Sci.* **2015**, *60*, 509–512. [CrossRef]
83. Liu, L.; Zeng, D.; Yang, M.; Wen, B.; Lai, J.; Zhou, Y.; Sun, H.; Xiong, L.; Wang, J.; Lin, Y.; et al. Probotic *Clostridium butyricum* improves the growth perfomance, immune function, and gut microbiota of weaning rex rabbits. *Probiotics Antimicrob. Proteins* **2019**, *11*, 1278–1292. [CrossRef]
84. Matur, E.; Eraslan, E. The impact of probiotics on the gastrointestinal physiology. In *New Advances in the Basic and Clinical Gastroenterology*; Brzozowski, T., Ed.; InTech: London, UK, 2013; pp. 51–74.
85. Dalle Zotte, A. Rabbit farming for meat purposes. *Anim. Front.* **2014**, *4*, 62–67. [CrossRef]
86. Dalle Zotte, A. Perception of rabbit meat quality and major factors influencing the rabbit carcass and meat quality. *Livest. Prod. Sci.* **2002**, *75*, 11–32. [CrossRef]
87. Li, R.G.; Wang, X.P.; Wang, C.Y.; Ma, M.W.; Li, F.C. Growth performance, meat quality and fatty acid metabolism response of growing meat rabbits to dietary linoleic acid. *Asian Astr. J. Anim. Sci.* **2012**, *25*, 1169–1177. [CrossRef] [PubMed]

88. Mattioli, S.; Cardinali, R.; Balzano, M.; Pacetti, D.; Castellini, C.; Dal Bosco, A.; Mattioli, N.G.F. Influence of dietary supplementation with prebiotic, oregano extract and vitamin E on fatty acid profile and oxidative status of rabbit meat. *J. Food Qual.* **2017**, *2017*, 3015120. [CrossRef]
89. Ayyat, M.S.; Al-Sagheer, A.A.; Abd El-Latif, K.M.; Khalil, B.A. Organic selenium, probiotics, and prebiotics effects on growth, blood biochemistry, and carcass traits of growing rabbits during summer and winter seasons. *Biol. Trace Elem. Res.* **2018**, *186*, 162–173. [CrossRef] [PubMed]
90. Shah, A.A.; Liu, Z.; Qian, C.; Wu, J.; Sultana, N.; Zhong, X. Potential effect of the microbial fermented feed utilization on physicochemical traits, antioxidant enzyme and trace mineral analysis in rabbit meat. *J. Anim. Physiol. Anim. Nutr.* **2020**, *104*, 767–775. [CrossRef]
91. Cummings, J.H.; Macfarlane, G.T.; Englyst, H.N. Prebiotic digestion and fermentation. *Am. J. Clin. Nutr.* **2001**, *73*, 415S–420S. [CrossRef]
92. Hermida, M.; Gonzalez, M.; Miranda, M.; Rodríguez-Otero, J.L. Mineral analysis in rabbit meat from Galicia (NW Spain). *Meat Sci.* **2006**, *73*, 35–639. [CrossRef]
93. Valenzuela, C.; Lopez de Romaña, D.; Schmiede, C.; Sol Morales, M.; Olivares, M.; Pizzaro, F. Total iron, heme iron, zinc and copper content in rabbit meat and viscera. *Biol. Trace Elem. Res.* **2011**, *143*, 1489–1496. [CrossRef] [PubMed]

MDPI
St. Alban-Anlage 66
4052 Basel
Switzerland
Tel. +41 61 683 77 34
Fax +41 61 302 89 18
www.mdpi.com

Animals Editorial Office
E-mail: animals@mdpi.com
www.mdpi.com/journal/animals

www.ingramcontent.com/pod-product-compliance
Lightning Source LLC
LaVergne TN
LVHW070612170726
843515LV00004B/58

* 9 7 8 3 0 3 6 5 0 9 5 4 9 *